The Biology of Rocky Shores

Biology of Habitats

Series editors: C. Little, M.H. Martin,
T.R.E. Southwood, and S. Ulfstrand.

The intention is to publish attractive texts giving an integrated overview of the design, physiology, ecology, and behaviour of the organisms in given habitats. Each book will provide information about the habitat and the types of organisms present, on practical aspects of working within the habitats and the sorts of studies which are possible, and will include a discussion of biodiversity and conservation needs. The series is intended for naturalists, students studying biological or environmental sciences, those beginning independent research, and biologists embarking on research in a new habitat.

The Biology of Rocky Shores

Colin Little and J. A. Kitching

OXFORD
UNIVERSITY PRESS

OXFORD
UNIVERSITY PRESS

Great Clarendon Street, Oxford OX2 6DP

Oxford University Press is a department of the University of Oxford.
It furthers the University's objective of excellence in research, scholarship,
and education by publishing worldwide in

Oxford New York

Athens Auckland Bangkok Bogotá Buenos Aires Cape Town
Chennai Dar es Salaam Delhi Florence Hong Kong Istanbul Karachi
Kolkata Kuala Lumpur Madrid Melbourne Mexico City Mumbai Nairobi
Paris São Paulo Shanghai Singapore Taipei Tokyo Toronto Warsaw

with associated companies in Berlin Ibadan

Oxford is a registered trade mark of Oxford University Press
in the UK and in certain other countries

Published in the United States
by Oxford University Press Inc., New York

A catalogue record for this book is available from the British Library

Library of Congress Cataloging in Publication Data
Little, Colin, 1939–
The biology of rocky shores / Colin Little and J. A. Kitching.
(Biology of habitats)
Includes bibliographical references and index.
1. Seashore biology. 2. Seashore ecology. I. Kitching, J. A.
II. Title. III. Series.
QH95.7.L57 1996 574.5'2638—dc20 95–42040
ISBN 0 19 854935 0 (Pbk)

Printed in Great Britain
on acid-free paper by
Bookcraft Ltd., Midsomer Norton, Avon

Preface

Contrary to the opinion of the Rat, in *The wind in the willows*, there is nothing half so much worth doing as simply messing about on a rocky shore (with or without a boat). In spite of this truth, so obvious to us, we could find no book that we felt would really convey to students the interest to be found there—so we presumptuously decided to write our own. It is based on many years of fieldwork on the shore, courses of lectures that we have given, and especially on our experience of running field courses. We hope it will encourage students to take up a science that provides enormous intellectual rewards, coupled with the pleasure of working in some of the last easily accessible places relatively unspoilt by man.

The book is traditional in organization, and leads from the study of individual species to the study of communities. We assume that readers will have some taxonomic knowledge, but in Appendix I we give a brief illustrated guide to genera to help those who find it hard to remember names. We discuss examples from around the world to show how field experiments can be used to investigate the processes that occur on the shore. Inevitably, some sites have been the focus of much attention, and their names keep cropping up. To help the reader locate these, there are maps in Appendix II.

In our experience we have found that most people develop an interest in rocky shores by beginning with individual species or groups of plants or animals. We have therefore provided some detailed information about particular species in north-west Europe. However, the reader will find that in the chapters devoted to groups of organisms, a general section first describes world-wide distribution, and another discusses the general biology of the group, with examples from various parts of the world. We hope, therefore, to place north-west Europe into a global context. At the end of most chapters are brief suggestions for experiments or observations that can be carried out either by classes or individuals.

We have had to draw upon an enormous body of literature, and would like to thank the authors of well over 600 papers for their contributions to our knowledge of rocky shores, although space has allowed us to quote relatively few of these. Many of the illustrations are based on published work, and we hope the original authors will be happy with our simplified versions, all of which have been completely redrawn.

We have been lucky to benefit from comments made by a series of referees at various stages. They have helped to shape the book into a form that we hope will prove useful. While imparting no blame to them for our mistakes, we would like to thank them. In particular, we would like to thank Dr Beth Okamura, who has commented on scientific content of the whole book; and Prof. T.R.E. Southwood, Dr Cathy Kennedy and Penny Stirling, who have attempted to make us present the book in a fashion that will be clear, yet keep readers awake. We also gratefully acknowledge help and advice about conservation and coastal protection from Trevor Lloyd of the Civil Protection Planning Unit, Dyfed, and Paul Gilliland of English Nature. Lastly, we thank Mrs Joyce Ablett for her labours in producing the index.

April 1995   Colin Little and J.A. Kitching

Contents

1 The shore environment: problems for organisms and their investigators

In July 1980, a survey of rocky shores in the Rance estuary, in northern France, found that the more marine areas were well populated with a variety of marine snails—limpets, topshells and winkles. At one site, the topshell *Monodonta lineata* was common at all tidal levels. Then, after a sudden change from dull wet weather to hot sunshine, most individuals from high on the shore became moribund and accumulated near the bottom of the shore. The population was devastated. Fifteen months later, in October 1981, monitoring work in the Severn estuary, in south-west Britain, showed the fauna of a rocky promontory to be dominated by the limpet *Patella vulgata*. But in January 1982, after a succession of cold days and a night when temperatures plummeted to −13°C, many of these limpets lost their ability to cling to the rock and could be picked up by hand. While they were probably not killed outright, they were knocked from the rock by waves, and the population suddenly declined.

These two casual observations make the point that while most organisms on rocky shores are fundamentally 'marine', they have to cope with being out of water at regular intervals, and so have to deal with the rigours of life in air. Without water cover, most algae stop photosynthesis, and most aquatic animals stop feeding. Respiratory problems arise for these animals because aquatic exchange systems work only very inefficiently in air. Temperature changes and desiccation cause major problems for both plants and animals. At high temperatures, enzyme systems can be damaged, and desiccation may damage external membranes as well as disrupting circulation and the water content of tissues. At low temperatures the most serious problems are caused by the physical effects of ice formation. On a small scale, ice formation within tissues can break cell membranes, while on a gross scale ice scour can remove whole communities from the rock surface. Marine organisms thus 'endure' periods of low tide while they wait for the water to return.

Most marine biologists, on the other hand, investigate shores at low tide, unless they are scuba divers, or have access to remote sensors such as underwater video cameras. There are many repercussions from this incompatibility of timing. The most obvious of these, perhaps, is that investigators rarely see organisms functioning naturally and interacting with others, so that while their experiments tell us the results of any interactions, they seldom tell us the mechanisms involved. Another repercussion is that recordings of the distribution of mobile fauna may be misleading because at low tide the animals are not foraging but have retreated to protective refuges. For marine biologists as well as for marine organisms, therefore, the very first physical property of the intertidal zone requiring attention is the rise and fall of the water level, and we begin this chapter by briefly discussing tides and waves.

The approach taken by different biologists when working on the shore varies immensely. Some workers may begin study by surveying the occurrence and distribution of particular species of plants and animals. Others may be interested in one taxonomic group, and examine population structures or behavioural interactions. Those with a physiological bent may study the ways in which organisms are adapted to their environment. Whichever approach is adopted, most studies will need some kind of physical background survey. They will also require accurate identification of the organisms being observed. Also, unless the survey is of the 'presence/absence' type, the observers will require methods for estimating the densities of organisms on the shore. We describe some of these techniques after our discussion of tides and waves.

Tides and the problems of being out of water

Tides are caused by the gravitational pull of the moon and the sun on the ocean waters of a rotating earth. Various astronomical features also determine the changing characteristics that are shown by the tides from day to day, month to month, and over the seasons. Briefly, we can summarize these as follows.

The daily tidal pattern

The tide is actually a wave—admittedly of very long wavelength (something of the order of 1000 km)—and water generally rises on the shore in the form of a modified sine curve. From low tide it rises slowly; then the rise quickens until at mid-tide it is rising at its fastest. Towards high tide, the rate of rise slackens off to zero.

The most common tidal pattern is a semidiurnal one: there are two high tides and two low tides each day. This is because the major tide-raising force comes from the moon, which rotates around the earth roughly once each 24 h. However, the exact period of rotation is 24.84 h, so that the tides are later (on average) by 0.84 h each day. But why are there *two* tides each day, not just one? An answer to this question requires an explanation of the forces acting on the moon and the earth (Fig. 1.1). Briefly, these can be summarized by saying that there is a gravitational attraction between the earth and the moon, which is balanced by centrifugal forces tending to throw the two bodies apart. These latter forces occur because the earth and the moon rotate about a common axis. Note that the point of rotation of this axis is *not* in the centre of the earth, so that centrifugal forces on the earth act on the side of the earth opposite to the moon. There they cause the sea to 'bulge', creating a high tide. Meanwhile, the gravitational forces on the side of the earth nearest to the moon cause a second, equal 'bulge'—another high tide. As the moon rotates around its common axis with the earth, these two tides move round the earth.

Fig. 1.1 Forces acting on the earth and the moon. Gravitational forces (G) oppose centrifugal forces (C). On average, these balance out; but on the earth's surface closest to the moon, G is greater than C, while on the opposite side of the earth C is greater than G. The earth therefore shows two tidal bulges.

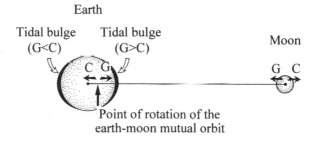

A simple pattern of two equal tides per day (with a delay of approximately 50 min from day to day) is therefore common, but by no means universal. The two tides can be of very different sizes. Partly this is due to the varying declination of the moon—the angle between the moon and a line through the earth's equator. Partly it is due to complexities caused by the flow of the tide wave through waters that are partly blocked by land masses. The overall variation runs from fully semidiurnal, through semidiurnal with unequal tides, to diurnal, where there is only one high and one low tide each day. As a result, organisms on shores in different parts of the world experience quite different patterns in the times when they are forced to tolerate life in air.

Changes in tides over the lunar month

While there may be differences in height between the two tides observed in one day, larger changes are usually to be seen over periods of weeks. These changes are determined by changes in the relative tide-raising forces generated by the sun and the moon (Fig. 1.2). When the gravitational attractions of sun and moon act together, i.e. when sun, moon and earth are in a straight line, tide-raising forces are at a maximum. This occurs at the times of new moon (when the sun and the moon are both on one side of the earth) and full moon (when the sun and the moon are on opposite sides of the earth). A few days after these occurrences, high tides are very high while low tides are very low, i.e. tidal amplitude is very large. These tides are called 'spring' tides, named from the sense of 'spring' meaning 'to rise' and not from the season, though as will be seen later, there are high spring tides in the spring. Marine biologists are most commonly seen on the shore at these times, because only then can they examine low tidal levels.

Fig. 1.2   The monthly lunar cycle and its relation to spring and neap tides. When sun, earth and moon are in line, gravitational forces add up, to give spring tides. When the moon is not in line, its gravitational attraction acts in opposition to that of the sun, giving neap tides.

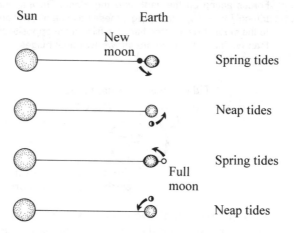

When the moon is at right angles to a line from the sun to the earth, the gravitational attractions of sun and moon act in opposition. Tide-raising forces on the earth are at a minimum, high tides are not very high, and low tides are not very low: tidal amplitude is small. These tides are called 'neap' tides from an Old English word 'nép' (meaning unknown).

Spring tides change gradually to neap tides over a period of two lunar weeks, so that a complete cycle of springs to neaps and back again takes one lunar month (29 days) (Fig. 1.3).

Changes in tides over the seasons

Not all spring tides are of the same amplitude. At the equinoxes (21 September and 21 March), the highest springs occur, while at the solstices (21 June and 21 December), the spring tides are at their lowest amplitude. Many shore workers run field courses near the equinoxes, when the lowest regions of the shore are uncovered. The seasonal differences are caused by the changing declination of the sun: at the equinoxes, the sun is overhead at the equator, so that sun, moon and earth are nearest to being in a straight line. At the solstices, the sun is overhead north or south of the equator, and the moon–earth–sun line deviates most from being straight.

Fig. 1.3 A lunar month, showing predicted tides for Plymouth, England. The monthly cycle contains two neap tide periods and two spring tide periods. Note also that alternate high tides are often of different heights. CD is chart datum (see p. 6). (From information in the Admiralty tide tables.)

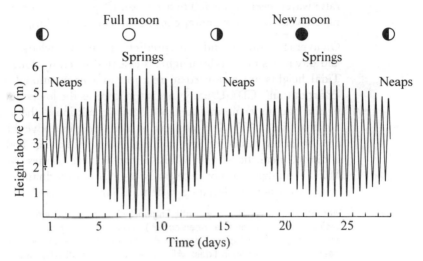

Tidal predictions and tidal terminology

Because of the variations in tidal amplitude and in the times of high and low tide just discussed, the pattern of tidal rise and fall at any one site is quite complicated. How can one predict this pattern, and how can one relate it to patterns seen at other sites?

Tidal predictions are available in various forms of 'tide table'. In these volumes, times and heights of tides are predicted for a number of standard ports, and corrections are given for a large number of secondary ports. Tidal amplitudes vary immensely around the world. For instance, the maximum tidal range in the Bay of Fundy, in Canada, is around 15 m, while in parts of the Mediterranean there is no appreciable tide at all. The times of tides also vary from place to place. Low water of spring tides in the Severn Estuary in Britain is always around the middle of the day and the middle of the night, while 300 km away on the Isle of Man it is always in early morning and late evening.

It is important to note that the figures given in tide tables are predictions, under standard meteorological conditions. Two types of change in the weather can affect these predictions. First, changes in barometric pressure may cause tides to be as much as 0.3 m different from predicted levels. High pressure depresses sea level while low pressure raises it, so that more of the low shore is uncovered during times of high barometric pressure. 'Fine-weather biologists' can therefore expect to see more of the rich diversity of low-shore biota than those choosing the theoretically 'best' tide on a day of rain. Secondly, the action of wind can pile water up against the shore, or drive it offshore, altering both tidal heights and times. Strong winds blowing along the coast may also cause 'storm surges', and these may have massive effects on water level. Surges that raise water level by 0.6–0.9 m are not infrequent, and some surges may raise levels by 3 m or more, causing disastrous flooding.

On average, however, tides are predictable, and knowledge of tide times and of various mean tidal heights is essential when working on a shore. Tidal heights are measured in relation to a conventional level called 'chart datum' (CD). Usually this is the same as the lowest level reached by the tide under normal meteorological conditions—the lowest astronomical tide (LAT). The most useful mean heights above this (Fig. 1.4) are the mean values for low and high water of springs and neaps. In addition, mean tide level (MTL) is the average of these four heights. MTL is similar (but not necessarily identical) to ordnance datum, the height to which the levelling system on land is referred.

Fig. 1.4   Predicted tidal curves for Plymouth, England, showing an extreme spring tide (solid circles), a mean spring tide (open circles) and a mean neap tide (triangles). EHWS, extreme high water of spring tides; MHWS, mean high water of spring tides; MHWN, mean high water of neap tides; MTL, mean tide level; MLWN, mean low water of neap tides; MLWS, mean low water of spring tides; ELWS, extreme low water of spring tides. (From information in the Admiralty tide tables.)

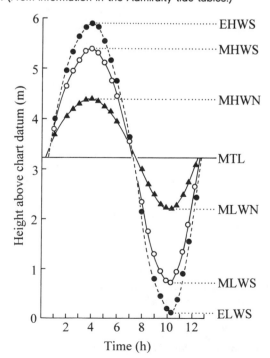

Changes in conditions over the tidal cycle

The rise and fall of the tide over the shore can be referred to as the 'emersion–submersion' cycle. The shore (and organisms on it) are 'emersed' as the tide falls, and this terminology avoids confusion with the word 'exposed' which is normally retained in the sense only of 'exposure to wave action' (see p. 8).

Despite the obvious importance to intertidal organisms of the radical changes in conditions that occur during emersion, there have not been many studies in which physical conditions on the shore have been examined in detail. Most workers have concentrated instead upon the tolerance of animals and plants to factors such as changes in temperature and to desiccation. From this approach it has perhaps been inevitable that such factors have often been seen as major direct controllers of distribution on the shore. There is, however, little evidence that climatic regimes on temperate rocky shores are normally important sources of mortality to animals (Little, 1990). In many cases physical conditions are only indirectly important because they influence processes such as competition and predation. This is not to say, however, that factors such as temperature are always unimportant. Some algae at the top of the shore, for instance, may be killed by high temperatures lasting for just a few crucial days in summer.

Nevertheless, the climate on the shore is quite complex. At Plymouth, temperatures in a variety of microhabitats were compared with the air temperature recorded at a nearby meteorological station (Southward, 1958). Air temperatures on the shore bore little relationship to the met. station temperature, and were usually higher by several degrees. It was difficult to measure rock temperatures directly, so measurements were made inside dead barnacle shells filled with plaster of Paris to give an idea of the temperature experienced by an inanimate body. These were always much higher than the air temperature in the shade, sometimes by as much as $10°C$. Temperatures of live barnacles were also higher than that of the air, but the barnacles were not always as warm as the inanimate bodies, suggesting some cooling by evaporation. The effect of sunlight in warming some of the animals was dramatic: on a day when it was $-2°C$ in the shade—low enough to freeze animals—the body temperature of limpets was around $8°C$. These observations serve to demonstrate the complexity of what might be called the 'effective' microclimate on the shore: when out of water, animals experience a range of conditions that cannot be predicted from meteorological readings.

Periodic emergence into air, or emersion, is an overwhelming feature affecting life on the shore. Consequently, much attention has been paid to the shape of the annual 'emersion curve'—a plot of how much time organisms spend out of water at various tidal levels. In some places, the rise and fall of the tide produces a smooth S-shaped 'sigmoid' emersion curve, but in others the complexities of local conditions lead to extremely irregular curves. Some curves have breaks or changes in slope at particular tidal levels, suggesting that physical conditions might change significantly at these heights. The relevance of such 'critical tidal levels' to the distribution of organisms on the shore is discussed on p. 19.

Waves and the problems of 'exposure'

Waves exert a destructive mechanical effect, encourage scour by sand and shingle, circulate water, disturb or deposit sediment, renew oxygen and reduce dissolved carbon dioxide. They also affect the movements of animals and thereby limit feeding or keep away predators, and can splash areas that would not normally be covered by the tides. The extent of wave action and its physical effects are, however, exceedingly difficult to measure. Most shore workers lump the overall effect of waves on the shore as 'exposure', but it is important to realize that this term contains a complex of factors.

The degree of wave action varies enormously between shores. In a few extremely sheltered sites, such as the upper reaches of some of the long narrow sea lochs in Scotland or the fjords in Norway, the sea surface is calm and the tide rises and falls as a still surface. Wherever there is some distance of offshore water, however, the action of the wind produces waves. In deep water, such waves may be enormous, yet individual 'particles' of water move in circular orbits without moving forward in the direction of the wave. When waves reach shallow water (a depth of less than one-twentieth of the wave length), the circular motions of the water particles are transformed into ellipses. At even shallower depths, particles move backwards and forwards so that they travel alternately onshore and offshore. At the same time, wave height increases while wavelength decreases, and at a critical point the wave form becomes unstable: the wave breaks (Pethick, 1984). The resulting wave action on rocky shores is a major factor influencing the type of habitat available for colonization by animals and plants, and we need to discuss how it can be measured before proceeding further.

Indirect ways of measuring exposure

The factor that influences the amount of wave action on a shore most is how far the wind can blow towards the shore over an uninterrupted stretch of sea. This distance is known as the 'fetch'. Many shores on the south-west coast of Ireland or in north-west France face the Atlantic Ocean and have fetches of 5000 km. Such shores are also open to winds blowing from several different directions (Fig. 1.5). In this case, the prevailing winds are from the south-west, ensuring that waves reaching the coast are both frequent and large. Conversely, shores in north-west Scotland at the heads of sea lochs, or in fjords in Norway, may have a fetch of only a few metres, and such shores are not open to prevailing winds from the south-west.

Fig. 1.5   Maps of two indented coastlines to show variations in exposure to wave action. The circles centred on four localities are divided into 22.5° sectors (the same sectors for which wind data are usually reported). Those sectors which are no more than 50% obstructed by land have been shaded black. The number of these sectors gives some guide to exposure, but other factors such as direction of the prevailing wind, and depth of offshore water, should be taken into account. Loch Eil in Scotland is extremely sheltered, while Mizen Head in Ireland is extremely exposed. (Partly after Thomas, 1986.)

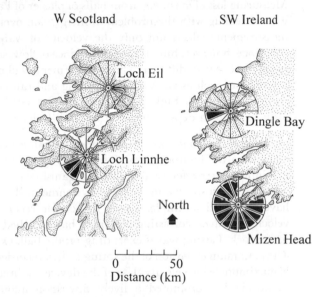

Various other factors besides fetch affect the wave 'climate' of a shore. The direction of the prevailing wind has already been mentioned. The slope of the shore is very important in controlling the type of waves and the point at which they break. On relatively flat shores, waves break far out from the shoreline, then 'spill' over the shore, while on steep ones the waves come close in before 'surging' up the rock face. Surging waves on steep shores may increase the effective intertidal zone to many times the height predicted purely from tidal rise and fall.

Using measurements of fetch, angle of the shore open to the sea, wind velocity and shoreline profile, one can calculate some form of 'exposure index' (Baker and Wolff, 1987). Such calculations involve a number of assumptions, and in any case it is difficult to be sure of which aspects of 'exposure' are the critical ones for littoral organisms. Nevertheless, exposure indices have the advantage that they can be calculated from data readily available in charts and weather records, and they can be very useful as comparative measures.

Field measurements of exposure

An alternative way of trying to estimate exposure is to measure some of the physical effects of wave action directly. This approach is not easy, because wave action is not continuous, and the effects of, say, one severe storm may be greater than those of the waves throughout an entire year. Measurement has, therefore, to integrate effects over long periods of time. There have been essentially two approaches. One has been to measure how fast some erodible substance wears away; the other has been to measure some aspect of water velocity.

Measuring loss of mass, e.g. from balls of plaster of Paris, is a very simple way of dealing with the problem of integration over time. This kind of measurement reflects not only the velocity of water moving past the substance, however, but also the turbulence of flow, so that calibration of the method is very difficult. The measurements will probably relate well to processes such as rates of filter feeding and rates of larval settlement (which depend on bulk water motion) and fertilization and passive dispersal (which depend on the rate of diffusion).

Other processes, such as the likelihood of being knocked off the rock, depend on the hydrodynamic force generated by the water. Thus the maximum water velocity to which an organism is exposed may determine whether it will survive in a given environment. Bell and Denny (1994) have described a simple, cheap device for recording the maximum velocity of waves, consisting of a small ball attached to a spring secured to the rock. Passing waves exert drag on the ball, extending the spring. The maximum extension of the spring is then recorded, giving a measure of maximum wave velocity. Use of the device on three shores in northern California has promoted a lively discussion about what constitutes 'exposure', because the highest velocities were recorded not on open coasts (classically regarded as most exposed), but on apparently less exposed shores where waves were 'funnelled' into small channels. This observation makes it difficult to explain the distribution of those organisms classically regarded as typical of 'exposed' shores, solely in terms of the maximum velocity of water motion. We return to this point, and to the overall effects of waves on shore organisms, on p. 35.

Shore topography and the problems of physical surveys

One of the major problems when working on the shore is to determine the vertical height at which observations are made. It is usually very important to know this because the height relates to coverage by the tides, and can therefore be related to many physical variables. To make comparable observations on a series of shores it is essential to make them at comparable heights. But what *are* comparable heights? Heights measured in relation to the land levelling system (ordnance datum, OD), for instance using bench marks, have different relations to chart datum (CD) at different sites. Any measurements on the shore made relative to OD must therefore be converted to CD, and tables are available for this in most tide tables. As mentioned above, OD is approximately equivalent to mean tide level. If fixed heights such as bench marks are not available, one can use the tides themselves: a number of recordings of tidal height at high tide or low tide in average weather conditions can be used to give a reasonable estimate of the height of CD.

Supposing that a reference height has been established, the next likely problem is to make some record of the vertical profile of the shore. This can be done in a number of ways, depending upon the detail required (Baker and Wolff, 1987). For great accuracy, a surveyor's or builder's level can be used. This consists essentially of a small telescope mounted on a tripod and levelled using a spirit level. The telescope is used to sight a 'staff' or graduated pole at a distance downshore. On most rocky shores, this apparatus is clumsy to use, and can be replaced by a 'cross-staff', which is cheap, portable and surprisingly accurate over short distances. Cross-staffs, as shown in Fig. 1.6, are used to determine points on the shore at set vertical increments, say 0.5 m, and are used by surveying up the shore.

Fig. 1.6 A cross-staff in use on the shore. When the observer adjusts the spirit level to horizontal, the line of sight along the level allows Station 2 to be fixed at a height *x* above Station 1.

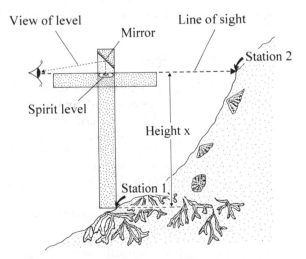

From heights measured by level, cross-staff, or a variety of other devices, and distance between points as measured by a tape, a crude profile of the shore can be drawn. If required, detail can be added by making detailed sketches or by measuring supplementary points.

Organisms and the problems of collecting, identifying and looking after them

Whether the approach on the shore is concerned with studying some aspect such as marine biodiversity, or with studying some specific research topic, the starting point is likely to involve qualitative collections of fauna and flora, and their identification.

Collecting on the shore

Collecting a representative set of organisms is mainly a matter of careful observation and of the examination of as many microhabitats as possible. Even an apparently uniform shore consists of very many microhabitats, including, for example, the undersides of boulders and overhangs; the insides of crevices; the surfaces of seaweeds; and, on a smaller scale, the insides of dead barnacle shells. Boulders and outcrops with different aspects will have different communities, while overall community structure will change in relation to vertical height and wave exposure.

Within this variety of habitats, it may also be important to have the right 'search image'. Many students will ignore organisms such as bryozoans, sponges and diatom films because they do not recognize them as living. Conversely, experienced workers may not be able to locate organisms that they have not seen before, purely because their expectations of how they should appear are not accurate. For all these reasons, a period of general observation of a shore is advisable before any detailed work is undertaken.

Looking after fauna and flora

When organisms are removed from their normal habitat on the shore, they can very easily be damaged, and may die rapidly unless they are treated carefully. Animals, in particular, need clean water with a high oxygen level. The oxygen problem is best dealt with by placing very few animals in any one container, and ensuring there is an air space twice as large as the volume of water in the container. More tolerant animals such as gastropods can be brought back in damp seaweed. On return to a laboratory, animals should be placed in containers of clean sea water as soon as possible. Delicate organisms such as echinoderms and fish require aeration.

After examination, material collected from the shore should be returned to the same shore from which it was collected. Dead material, since it will contribute to the detritus pool, should also be taken back to the shore.

Identification of organisms

There are perhaps three tiers of guides to the organisms found on rocky shores, their use depending upon the expertise of the investigators. General texts can be used to place an organism in a phylum or class, or sometimes into a family. More specific 'marine' texts, appropriate only for smaller geographical areas, may be straightforward identification guides, but some of the regional accounts of shores have become classics of marine biology. For ultimate detail, specialized guides are necessary, and for many taxa reference must be made to specific published papers. We give some guidance about identification texts in the list of further reading (p. 215).

One particular problem facing an investigator when trying to identify an organism may be mentioned here. Scientific names tend to change over the years—perhaps because an earlier description of the type species is discovered to have priority, or perhaps because it is recognized that a 'species' really consists of more than one taxon. An example of the latter is found in the barnacles, where *Chthamalus stellatus* has been subdivided into *C.stellatus* and *C.montagui*. Unfortunately this means that the species referred to in early papers is not clear. The status of *Littorina* snails is even more complex. '*Littorina saxatilis*' has suffered many divisions and recombinations, and in this case very few texts can be entirely up to date so that it may be necessary to seek help from specialists.

Making quantitative observations

It is seldom possible to observe, or even to count, every individual of any particular species on the shore. All estimates of populations therefore tend to be made from counts of smaller samples, and subsequent statistical analysis. We cannot deal here with details of statistical procedures, but we will attempt to give some introduction to the ways in which quantitative observations can be made. A good discussion is given by Baker and Wolff (1987).

In general, the use of quadrats has overtaken all other measures: the observer counts organisms within a number of quadrats, and then estimates the population size or density from a mean value. Three main questions will plague the investigator using quadrats: 'What size should they be?', 'How many counts are necessary?' and 'How should the quadrats be distributed?'

Quadrat size must be chosen in relation to the size and density of organisms. For instance, a 10 cm × 10 cm quadrat, subdivided into 1 cm × 1 cm squares, is useful in counting the smaller barnacle species; whereas a 50 cm × 50 cm quadrat may be more appropriate for limpets. For algae, individual plants cannot usually be counted, and some kind of 'percentage cover' estimate is more usual. The simplest way of doing this is to use a 50 cm × 50 cm quadrat which has been strung with wires crossing at right angles, forming 100 'points'. Individual points lying above a particular algal species, when added up, give percentage cover.

When using quadrats, it is always difficult to judge how many measurements to make to give a reliable mean. One of the simplest approaches is to plot a 'running mean', in which the mean is recalculated each time another measurement is added to the total. When the mean is plotted against the number of samples, and the standard deviation is added as an error bar, the mean will be seen to stabilize eventually, suggesting how many readings are necessary.

It is also vitally important to understand the various ways in which quadrats can be laid out, and the implications these have for subsequent statistical analysis. Haphazard quadrats are simply positioned by throwing blindly over one's shoulder. To ensure that quadrats are truly random in position requires a more sophisticated approach. Using random number tables, quadrats can be placed either at random along a line, or within a grid. For continuous monitoring surveys, quadrats fixed in position may be more useful (but see the problem of pseudoreplication, below). There are occasions when quadrats laid out in some fixed pattern are better suited to analysis. Standard statistical texts should be consulted.

The use of sufficient replicates is essential in both survey work and in field experiments. In particular, attention must be paid to a problem known as 'pseudoreplication'. If replicate measurements are taken from only a small area of the shore, the results will refer only to that small area, no matter how many replicates are taken. Replicates should therefore be taken over a scale large enough to demonstrate that the phenomenon under investigation is of sufficiently widespread occurrence to be of general interest. The same problem may apply to measurements taken over time: it may be better to take measurements of different members of the population rather than to measure the same individuals again and again. For discussion, see Underwood (1986).

Strictly quantitative measurements can be extremely time-consuming, and a semi-quantitative method has now achieved widespread usage. This method employs 'abundance scales', in which densities are allotted to a category, say $1-10/m^2$, rather than being counted exactly. This method can considerably speed up the process of surveying a shore. An example, for algae, might have categories such as 'rare' (only one or two plants present); 'occasional' (scattered plants, zone indistinct); 'frequent' (less than 5% cover but zone still apparent); 'common' (5–29% cover), etc. In an example for animals the categories could be defined solely in terms of density, such as 'rare' (less than $1/m^2$); 'occasional' ($1-4/m^2$), etc. Many of these categories are easily distinguished by eye, but unfortunately the results cannot be analysed using normal statistics because the categories are actually ranks, not arithmetical units. They can, however, be processed using non-parametric techniques.

Safety and conservation on the shore

Working on the shore carries dangers both for observers and for the ecosystem itself. Here we make some points about precautions that should be taken.

Safety: a code of practice

All organizations should have a code of practice detailing safety features appropriate for the type of work being carried out. This may cover specialist techniques, such as scuba diving or the use of electrical appliances on the shore, as well as general precautions.

Safety: some general points about equipment

The most basic essential is appropriate clothing and footwear, which should be warm and waterproof—it is surprising how much colder it can be on the shore than on nearby mainland. In addition, a first-aid kit should be carried by the leader of the party, who must have some training in simple first aid. A rescue rope, preferably one that is designed to be thrown easily, should also be carried.

Safety: conduct on the shore

Rocky shores are often hazardous places to work because of the uneven terrain and the slippery surfaces produced by macroalgae and diatom films. These problems necessitate care in moving around. It is therefore always sensible to work in pairs.

Work near low water may have extra hazards from waves. On exposed shores, it may be necessary to use safety lines. Even when these are not considered necessary, one person in the party should watch for excessive waves.

Conservation: avoiding damage to the shore

Conservation is discussed in more detail in Chapter 9, but here it must be stressed that many intertidal marine ecosystems are in some ways quite fragile, despite their ability to withstand factors such as battering waves and long periods of emersion. It is therefore essential to cause as little damage as possible when visiting the shore. In particular, boulders must be returned to their original position after examination; and the temptation to collect many specimens of the same species must be resisted. Lack of attention to these points can seriously reduce species diversity on a shore, and may eliminate some types of community completely.

2 Vertical distribution: 'zonation' and its causes

On shores which have similar exposure to wave action, the distribution patterns of organisms are remarkably similar around the world. Almost everywhere, plants and animals are found in distinct bands or 'zones' at particular vertical heights and are not distributed randomly. Even a cursory visit to the shore will probably suggest this 'zonation' as the most obvious overall feature. For marine biologists, the zones are extremely useful in a practical sense because they provide a convenient descriptive framework into which specific observations can be fitted. But why do the zones exist? In this chapter we begin by describing zonation patterns found on moderately exposed shores, and then we go on to discuss the factors that establish and maintain them. We discuss the immense variation found on sheltered and exposed shores in Chapter 3.

'Zonation' on moderately exposed shores

Shores which are neither extremely sheltered nor pounded by extreme wave action usually have three distinct vertical zones (Fig. 2.1). The high-shore zone is dominated by small snails (e.g. *Littorina* spp.) and black lichens, together with blue-green algae, often extending well above the levels reached by tidal cover. Below this is a mid-shore zone characterized by barnacles and containing limpets, mussels and algae such as, in the north Atlantic, the fucoids. At the bottom of the shore is a zone uncovered only at low water of spring tides, containing red algae, laminarian algae (kelps), or (in parts of the southern hemisphere) large tunicates.

There has been much discussion concerning the terminology appropriate for these three zones, but we follow Lewis (1955) in calling the top zone the 'littoral fringe', the middle zone the 'eulittoral zone', and the bottom zone the 'sublittoral zone', which extends gradually downwards to regions well below those that are ever uncovered by the tides.

From Fig. 2.1 it is probably obvious that trying to split a shore into three zones is an oversimplification. There are many species within one zone, and some species that straddle zones. These complications are easily seen if we look in more detail at one particular shore (Fig. 2.2). Disregarding, for the moment, the two curves, the vertical bars show that both animals and plants have overlapping distributions, and that few species have exactly the same vertical distribution. This is the point at which one realizes that the idea of zonation has a limited usefulness: we need to be aware of its limitations.

Fig. 2.1   Diagrammatic view of a moderately exposed shore (Grade 4–5 on Ballantine's scale, see pp. 31–2) in the south-west of Britain, showing the pattern of zonation. Organisms are not drawn to scale, and only selected species are shown. There would also be, in particular, *Fucus vesiculosus*, red algae, topshells and dog whelks.

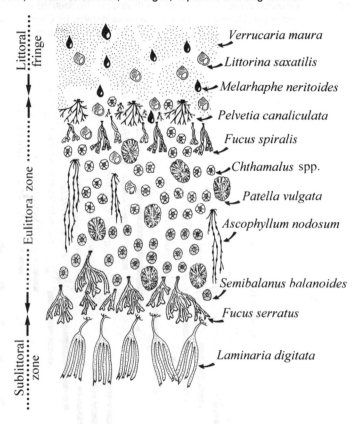

The first point to make is that we have defined the three zones entirely in biological terms. This is necessary because, although they are related to tidal levels, many other factors influence their heights—in particular the degree of wave exposure, as will be discussed in Chapter 3. Secondly, the zones are defined by using conspicuous organisms as indicators, but these are often very far from being evenly distributed across the zone. For example, the eulittoral zone in north-west Europe may have clumps of the alga *Fucus vesiculosus* which form an irregular and changing mosaic among a background of barnacles. Often, also, the zones have been deemed to divide readily into 'subzones'. In parts of southern Africa, for instance, the eulittoral zone has an upper subzone dominated by limpets, while the lower subzone is covered by an algal turf. Thirdly, the indicator organisms do not always appear in the same relationship: the barnacle line might end either above or below the fucoid alga *Pelvetia canaliculata*, especially if there is a change in barnacle species from *Chthamalus stellatus* to *Semibalanus balanoides*. Fourthly, the boundaries between zones are not necessarily sudden ones, but can occupy a transition belt. In particular, barnacles and black lichens such as *Verrucaria* might overlap considerably.

Fig. 2.2   Emersion curves for Plymouth, with vertical distributions of some of the gastropods and algae. Open circles show the curve derived by Colman; solid circles show that derived by Underwood. (After Colman, 1933; and Underwood, 1978.)

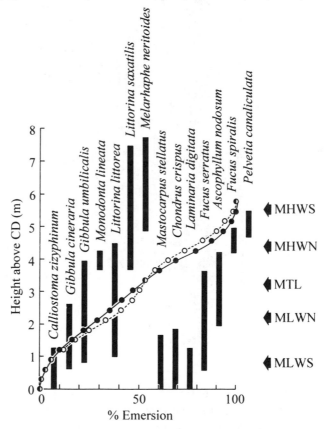

Even with these reservations, the three zones often provide a striking visual phenomenon: a clean-cut line may separate the upper limit of white barnacles below the black *Verrucaria* lichens. To see kelp beds appear as the water level sinks from MLWS to ELWS is to see the emergence of a 'forest' community with a multitude of component species not previously visible. Yet there have recently been suggestions that the zones may not be as 'real' as they appear. Because most zones are defined by qualitative reference to indicator species, e.g. 'the barnacle zone', there is little evidence that the zones really contain sets of co-occurring species that differ from those in the zones higher up or lower down. In other words, zones could simply be the visible overlapping patterns of distribution of individual, abundant and conspicuous species, such as barnacles, laminarian algae or lichens (Underwood and Kennelly, 1990).

Even if 'zones' are really only the visible patterns shown by individual dominant species, we are left with the problem of understanding why individual species are distributed in such well-defined vertical bands. Why do laminarians not occur further up the shore than they do? Why are barnacles restricted to a 'zone' between the laminarians and the high-shore lichens, and why do they often have a very sharp edge to their distribution? For it remains evident that as *descriptive* terms, the idea of 'zones' on the shore is a very useful one, particularly when comparing different shores.

We go on now to discuss the factors that may control the vertical distribution of organisms, and may therefore be said to control the extent and composition of the visible zones on the shore.

The influence of physical factors on zonation

Because the majority of organisms on the shore are marine in origin, it was natural that early investigators assumed zonation patterns to be determined by factors related to the length of emersion. These factors, such as desiccation and temperature changes, were assumed to limit the ability of organisms to live higher up the shore. Such assumptions seemed particularly reasonable because, when the tolerances of animals and plants were measured, those living higher on the shore were generally shown to have higher tolerances to being dried or heated than those living lower down. As we shall discuss later, however, this finding merely reflects the fact that organisms are usually well adapted to their environment. A complication, too, was that a theory based on tolerance of physical extremes did not obviously account for sudden changes in community composition at particular heights.

Critical tidal levels

One of the ways to evaluate changes in physical conditions up and down the shore is to plot an 'emersion curve' as already mentioned (p. 7). This curve shows the annual mean percentage time out of water for various heights on the shore, calculated from predicted values in the tide tables. When such a curve was plotted for Plymouth by Colman (1933), it showed, instead of the simple sigmoid shape that might have been predicted, a double sigmoid form (see the dotted line in Fig. 2.2). Colman also found that the vertical distribution of groups of species stretched between particular heights on this curve, which he called 'critical levels'. In later years, it was suggested that these critical levels occurred at the heights where there was a change in slope of the emersion curve, i.e. where there were sudden changes in the rate of increase or decrease in the total annual emersion time.

The facts of the case have been complicated by later comments on the shape of the emersion curve at Plymouth. Recalculations of the data from the tide tables have obtained a simple sigmoid curve, not a double one (see the solid line in Fig. 2.2); while the use of observed instead of predicted data again showed a double sigmoid curve. Emersion curves for some other shores in Britain, based on observations not predictions, have also shown two regions in which there are rapid changes of slope. In general, such changes are found in the regions of MLWN and MHWN, but it must be said that the deviations from a simple sigmoid curve are quite small. However, in these regions conditions do change drastically even if the changes do not appear in the annual emersion curve: below MLWN or above MHWN respectively, periods of *continuous* submersion or emersion are experienced at some times.

The concept of critical tidal levels has now been investigated experimentally (Underwood, 1978). To test the hypothesis that such levels existed, Underwood examined the vertical limits of distribution of common species on five shores in Britain. He used many (9–20) replicate transects at each site, and recorded the upper and lower boundaries of 5–28 species. On none of the five shores could he find evidence that the upper or lower boundaries of these species were distributed other than at random—in other words, he found no evidence that the limits of distribution of species were clumped at any particular levels. There is thus no good evidence for the existence of critical tidal levels in Britain.

However, in some areas of the Pacific coast of North America, the algae do seem to be distributed in such a way as to support the idea of critical tidal levels. Here there are very abrupt changes in the amount of emersion experienced at different heights on the shore because of a tidal regime in which alternate tides have very unequal heights. It seems, in summary, that critical tidal levels are not widespread, but that they can occur in aberrant tidal conditions. In general, the deviations of the emersion curve from a simple sigmoid shape are too minor to have any great physical significance.

This conclusion does not, however, mean that physical factors are without any effect on the vertical distribution of intertidal organisms, and we go on to consider their influences.

Direct physical effects

The overall effects of differing emersion periods can be examined experimentally by keeping species in aquaria which are filled with sea water and emptied for differing proportions of the time. This technique was used by Baker (1909) to investigate growth and survival of fucoid algae. When emersion was kept to 1 h in 12, the low-shore *Fucus serratus* grew very quickly, far outstripping all other species in the first 6–7 days. When emersion was increased to 6 h in 12, all species grew quite well, and it was here that *Fucus spiralis* (from high on the shore) grew best. In jars that were emersed for 11 h in 12, *Fucus serratus*, *Fucus vesiculosus* and *Ascophyllum nodosum* showed little growth and nearly all died in 2–3 days. All the *Fucus spiralis* plants survived, though they grew only slowly. Evidently the species at the top of the shore (such as *Fucus spiralis*) can tolerate loss of water but grow only slowly. Less-tolerant species, if given sufficient wetting, can grow faster and can therefore compete more successfully. Thus upward limitation may be controlled by physical factors such as desiccation, while downward limitation may be controlled by biotic factors such as growth rates. This conclusion is now accepted for a number of species.

Time out of water is not, itself, the only factor that limits the upward penetration of fucoid algae (Schonbeck and Norton, 1978). The uppermost plants of *Fucus spiralis* on the Isle of Cumbrae, Scotland, showed evidence of damage after sunny weather had coincided with neap tides so that the plants were continuously emersed for several days. This happened on five occasions between May and September 1975. After such an occasion in August 1975, *Ascophyllum nodosum* and *Pelvetia canaliculata* were also damaged. Thus, a few critical days in summer were sufficient to determine the upper limit of fucoid zones.

Another physical influence on vertical zonation of algae may be some characteristic of the change of light regime with depth. There has long been a theory that spectral composition of the light at various depths could be very important (Dring, 1982). Long wavelengths (red and infrared) are absorbed in the top metres of water, as are very short wavelengths (ultraviolet); so that light penetrating to below 10 m is mainly blue-green. One hypothesis proposed that the zonation of algae was related to their pigment type, so that blue-green and green algae were found in the upper intertidal, where chlorophyll was most efficient at absorbing light over a wide range of wavelengths. Brown algae occurred further down, using fucoxanthin pigments where red light was scarce. Red algae were found deeper, where phycoerythrin pigments were most efficient at absorbing the greener light. There is, however, little experimental evidence to support this idea. All marine plants near the sea surface in fact receive more light than they require, and therefore need protection from too much light at any wavelength rather than efficient mechanisms for absorbing it. It is now thought that pigment composition may be important in coping with differences in the total amount of light received at different depths.

Investigation of physical factors affecting animals on the shore has concentrated upon tolerance of desiccation and high temperatures, as discussed briefly in Chapter 1 (Newell, 1979). One example gives the flavour of this approach. Thermal resistance of the four common species of intertidal topshells from Plymouth was investigated in the laboratory by recording the time required, at a range of temperatures, to reach 50% mortality. The relation between lethal temperature and survival time was approximately a straight line when time was plotted on a logarithmic scale (Fig. 2.3). The order of tolerance in the laboratory parallelled the order of zones on the shore in which the species are usually found: *Monodonta lineata* is found in the upper half of the barnacle zone (the eulittoral); *Gibbula umbilicalis* occurs more towards the bottom of the

Fig. 2.3   Thermal resistance (in air) of four species of topshell. Solid circles show *Monodonta lineata* (*Ml*); open circles show *Gibbula umbilicalis* (*Gu*); solid triangles show *Gibbula cineraria* (*Gc*); and open triangles show *Calliostoma zizyphinum* (*Cz*). Resistance to heat stress follows the order in which the species are zoned on the shore, but there is little evidence to show that the species' tolerance causes their vertical distribution. (After Newell, 1979.)

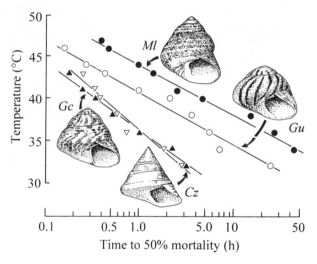

eulittoral; while *Gibbula cineraria* and *Calliostoma zizyphinum* belong to the sublittoral fringe. Could it be that the species are limited to their particular zones by their temperature tolerances and maybe by their temperature preferences? In reality, several points make this seem unlikely. First, the thermal tolerance of the snails much exceeded that necessary to survive normal temperatures on the shore. *Monodonta*, for example, could survive 40°C for 5 h. Secondly, the behaviour of the snails must be taken into account. *Monodonta*, at least, undertakes migrations up and down the shore over the seasons—yet it moves upshore in summer, suggesting that thermal stress is hardly likely to be limiting its distribution at Plymouth. Of course, temperature exerts sublethal effects on intertidal animals, as well as affecting mortality directly, and we have already noted

(p. 1) that *Monodonta* in northern France may be affected by summer temperatures. But it now seems likely that to regard physical factors such as high temperature as the normal direct determinants of vertical zonation patterns is much too simplistic a view.

The influences of biological interactions on zonation

In addition to considering the physical conditions on the shore, any theory of how vertical zonation is controlled must consider the biological interactions that occur between species. One way of observing how animals react to the tidal cycle is to conduct experiments in an artificial tide model—a laboratory device that mimicks the rise and fall of the tides (Underwood, 1972). Most mobile gastropods, when placed into this model, accumulated at the time of low tide in approximately the same positions that they would have occupied on a natural shore—allowing, that is, for the reduction of scale in the model. The winkle *Littorina obtusata*, however, behaved strangely. It crawled to the top of the artificial shore, above the high tide mark, and stayed there until it died from desiccation. Only when a band of fucoid seaweed was placed into the model did the winkles collect into one zone on the shore; and then they moved into the alga wherever it was placed. An attraction of the snails to the alga was therefore a crucial part of the zonation mechanism. Many more biological interactions are considered later in the book, in particular in Chapter 8. Here we discuss some examples that refer specifically to influences upon zonation patterns.

Larval settlement

The majority of intertidal organisms reproduce by means of 'propagules' that disperse in the plankton and thus have to recolonize the shore in each new generation. For algae the propagules are usually spores or gametes, while invertebrates have larvae that may pass through several stages of their life history in the planktonic phase. How do these dispersive phases come to form discrete bands on the shore? For mobile animals, the behaviour of individual adults is important (see p. 27), but for the majority of organisms which are sessile, settlement behaviour of the propagules may be an important factor. Since the spores of algae usually settle within relatively short distances of the parent plant (Chapter 4), and do not apparently show selection of height on the shore, we discuss here only the behaviour of invertebrate larvae.

The larvae of suspension feeders such as barnacles and spirorbid worms are extremely selective in the substrates to which they will attach, as we discuss further in Chapter 6. Some barnacles, in particular, settle at specific levels on the shore. This settlement behaviour accounts for a large part of the differences in zonation seen in *Semibalanus balanoides* and *Chthamalus stellatus*, two species common in north-west Europe. Larvae of the former settle from just above MLWS to just below MHWS, and it is therefore possible for adults to be found distributed throughout the eulittoral zone and the littoral fringe. Larvae of *Chthamalus*, in contrast, settle only between MTL and MHWS, so that adults are, by definition, restricted to the upper part of the shore.

In south-east Australia, two common barnacle species are *Tesseropora rosea* and *Tetraclitella purpurascens*. The larvae of neither species settles in the littoral fringe (Denley and Underwood, 1979) and the absence of adults at high shore levels can therefore be attributed to the behaviour of the larvae. Neither species settles where the substrate is covered by either algae or the tube-worm *Galeolaria*. Since the lower part of the eulittoral zone is dominated by macroalgae and *Galeolaria*, the absence of barnacles there can also be attributed to larval behaviour, especially since experiments with cleared substrates in these lower zones showed that larvae would settle there on appropriately clean surfaces.

Overall, the 'supply' of larvae to particular zones can be important in determining community structure. The term 'supply-side ecology' draws attention to the importance of factors that determine how available the propagules of individual species are at any site, and we discuss this further on p. 177.

Competition

Although the early experiments on growth rates in algae suggested that vertical distributions can be affected by competition, proof of this was not forthcoming until 'weeding' experiments were carried out on the shore (Norton, 1985). Areas in the *Fucus spiralis* zone on the Isle of Cumbrae, Scotland, were cleared while both this species and *Pelvetia canaliculata* in the zone above were fertile. Both species colonized the patches as sporelings, but those of *Fucus spiralis* were weeded out by hand. Under these circumstances, *Pelvetia* plants grew well, ranging down into the *Fucus spiralis* zone, showing that *Pelvetia* had previously been prevented from doing so by competition with *Fucus spiralis*. Under normal conditions, *Fucus spiralis* out-competed *Pelvetia* because of its immensely faster growth rate (Fig. 2.4).

Similar experiments on the Isle of Man showed that several fucoids could extend their range both up and down the shore if their competitors were removed. Thus when the *Fucus vesiculosus* above a zone of *Fucus serratus*, and the *Laminaria digitata* below it were removed, the *Fucus serratus* was able to expand both upshore and downshore. *Fucus serratus* was presumably prevented from extending downwards by the more vigorous growth of the alga below it; while it was prevented from extending upwards by the faster growth of the alga above it. It must be remembered, however, that as we have already seen, final upward limits of some fucoids may be set by physical factors.

Fig. 2.4   Growth of *Fucus spiralis* and *Pelvetia canaliculata* in the *Fucus spiralis* zone of the Isle of Cumbrae, Scotland, over a period of 9 months. Solid triangles show *F. spiralis*. Open circles show *Pelvetia* in undisturbed areas. Solid circles show *Pelvetia* in areas from which *F. spiralis* has been removed. *Pelvetia* grows much more slowly than *F. spiralis*, even when its competitor is removed. (After Schonbeck and Norton, 1980.)

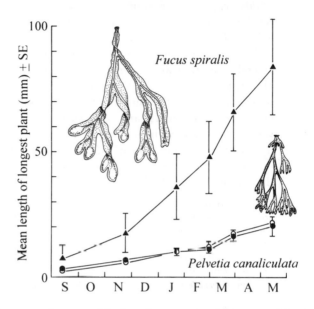

As will be seen when considering the structure of communities (p. 167), competition on the shore has major effects on the vertical distribution of the fauna. In the case of barnacles, *Chthamalus stellatus* and *Semibalanus balanoides* compete for space on the rock surface, and this (among other factors) leads to dominance by *Chthamalus* high on the shore but dominance by *Semibalanus* lower down. Limpets, too, show intense interspecific competition which affects vertical distribution. In many cases, competition results in a sharpening of the boundaries between zones so that species distributions hardly overlap.

Predation

The effects of predation have often been shown by using 'exclusion cages', made of a mesh fine enough to keep predators out. This technique was used on the Isle of Cumbrae, in Scotland, to examine the effect of the dog whelk *Nucella lapillus* on the vertical distribution of barnacles (Connell, 1961*b*). *Nucella* was found commonly only below MTL, and in the lower part of the eulittoral this whelk was an important cause of death in *Semibalanus balanoides*. However, the effects of *Nucella* were more complex than this simple result would suggest: predation by *Nucella* reduced barnacle densities so much that it reduced competition between *S. balanoides* and *Chthamalus stellatus*. This experiment showed well that biotic interactions can seldom be considered in isolation.

On the Californian coast, wave action was too severe to allow the use of cages, so the effects of a predatory starfish on the vertical distribution of a mussel were investigated by removing the starfish by hand (Paine, 1974). The starfish *Pisaster ochraceus* migrated upshore on to beds of the mussel *Mytilus californianus* in summer and autumn, eating large quantities of the bivalves. Normally the mussels formed a dense band in the upper eulittoral zone, about 1 m in vertical extent, and with sharp upper and lower boundaries. When the starfish were removed the mussel band extended dramatically downshore (Fig. 2.5), in the first year by 0.7 m. During the next 4 years, the limit extended down to a total of 0.9 m below its original level, so the mussel band nearly doubled in vertical extent. No changes were seen in a nearby control area. After 5 years, *Pisaster* removal was stopped, and in the next 3 years the lower limit of *Mytilus* retreated upshore again by about 0.6 m. *Mytilus* could thus form a dense zone in the upper eulittoral only because *Pisaster* rarely moved up to this level, and the bottom of the *Mytilus* zone was defined by *Pisaster* predation.

Fig. 2.5   The effect of removing the starfish *Pisaster ochraceus* (Po) on the vertical distribution of fauna at Mukkaw Bay, Washington State. Starfish were removed from spring 1963 to summer 1968 (see lines shown by arrows). *Mytilus californianus* (Mc) colonized downshore rapidly, accompanied by the stalked barnacle *Pollicipes polymerus* (Pp) (solid circles and line), but retreated upshore when *Pisaster* returned. Where *Pisaster* was present, a community dominated by *Balanus glandula* (Bg) and *Collisella digitalis* (Cd) prevailed (stipple). (From data of Paine, 1974.)

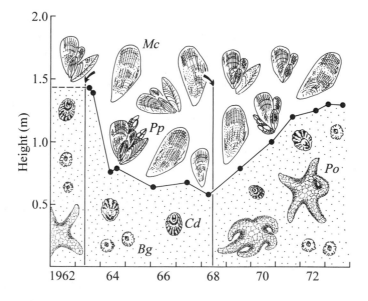

Grazing

Early experiments in which grazing limpets were removed from a strip of shore on the Isle of Man and observations of the aftermath of oil spills such as that from the *Torrey Canyon* (p. 82), which killed all species of grazers, showed the enormous effect that grazers can have on community composition. In Australia, grazing molluscs play a large part in eliminating macroalgae from the mid shore, although these algae are also limited in their upward colonization by desiccation (p. 181).

Experiments on the island of Helgoland, north Germany (Janke, 1990), have investigated the role of grazing winkles (*Littorina*) in determining zonation patterns on a very sheltered shore. Here the eulittoral zone is dominated by beds of the mussel *Mytilus edulis* and by algae (*Fucus* species), and the major grazer is the winkle *Littorina littorea*. On the low shore, there is a dense canopy of *Fucus serratus*, and *Littorina littorea* is accompanied by *Littorina mariae*. Janke carried out experiments in which both *Littorina* species were excluded from small areas. Although these experiments were not replicated, comparison with controls suggested that, on the mid shore, *Littorina* grazing excluded the green alga *Ulva* and reduced the settlement and growth of *Fucus* species. On the low shore, however, *Littorina* at its normal population density could not control the settlement and growth of *Fucus serratus*. Only when grazer numbers were doubled did they affect the *Fucus* canopy. Herbivory by *L. littorea* was therefore a major factor reducing algal cover in the mid shore, hence allowing the settlement of *Mytilus*, and giving the shore its characteristic zonation pattern.

Behaviour

Despite their mobility, gastropods such as winkles, and crustaceans such as crabs, usually maintain a well-defined zonation pattern. How do they achieve this? The first point to note is that the zonation patterns of these mobile animals seen on the shore at low tide do not represent the whole story. When covered by high tide, many species roam widely, as seen both in the field and in laboratory tide models. The crab *Carcinus maenas*, for instance, moves upshore as the tide rises, but retreats again as the tide falls. Some species of snails begin to move downshore as soon as the tide covers them, but return to their 'normal' position when the tide rises again.

Given that they move about, how do such species stay in the 'correct' zone? Some species, especially grazing gastropods, have been shown to exhibit movements that are random in direction and extent, at least in the short term (Chapman and Underwood, 1992). But since these animals usually remain within a relatively restricted zone, there must in the long term be some limits to movement in the vertical plane; and the very mobile species such as crabs evidently show directional movements. Field experiments on several species have shown that grazing gastropods displaced from their normal position are often capable of returning to it. When the topshell *Tegula funebralis*, common on the Pacific coast of North America, was moved out of rock pools to areas above, it returned to normal levels within a week. How do snails carry out such movements?

Though the answer is not known, there are several possible mechanisms. Snails can, for example, detect small changes in slope, presumably via their statocysts, and they respond to these in the laboratory. Their eyes are also capable of detecting some characteristics of the shore, at least the difference between light and dark areas. The topshell *Gibbula umbilicalis*, displaced seawards from its normal position in the field, returned shorewards towards a dark weed zone, and this movement could be reversed by positioning an artificial dark wall offshore. The chemical senses of gastropods are also acute, and several species have been shown to respond to the exudates of macroalgae in the laboratory. Although the cues from algal beds on the shore may be very diffuse, they may at least help snails to remain in the 'correct' zone.

Many mobile animals have been shown to possess endogenous activity rhythms, particularly those that trigger activity at each high tide. The crab *Carcinus maenas*, for instance, exhibits activity every 12.4 h when maintained in constant conditions. These circatidal rhythms may to some extent help to maintain the animals in the 'correct' zone because they switch off activity before the tide falls: they therefore reduce the possibility of being stranded above the normal resting level on the shore.

Overall, however, it must be said that there is as yet no coherent explanation of exactly how these mobile animals manage to roam up and down the shore but come to rest in particular zones at the time of low tide. This is a field that has been explored more thoroughly on sandy shores, but is still without resolution.

What causes zonation? The relative influences of biological and physical factors

Evidently both physical and biological factors are important in generating and maintaining zonation patterns on the shore. But what is their relative importance? It is difficult to generalize, because the number of factors involved is often high, and each interacts with the others. It does seem that in many cases intertidal species are controlled by some physical factors at their upper limit of distribution, while they tend to be controlled by sets of biological factors at their lower limit. A good example might be the distribution of the barnacle *Chthamalus stellatus*: the upper limit is set by tolerance of desiccation, while the lower limit is set by competition with *Semibalanus balanoides* and predation by the dog whelk *Nucella lapillus*. Similarly, the alga *Pelvetia canaliculata* is limited from living higher on the shore by desiccation, but is prevented from colonizing lower levels by competition from other fucoids. It seems unlikely, however, that the situation is often this simple; and as we have seen, experiments on some fucoids have shown that they have the capacity to withstand physical conditions higher on the shore than they are normally found, if other species are removed. Seaweeds also have to cope with grazing pressure from mobile gastropods, and we will show in later chapters that this varies from shore to shore depending upon degree of wave exposure.

Another generalization that has been made is that zones may be very broadly determined by physical factors, but that the boundaries of the zones are sharpened by biological interactions. Thus while the initial settlement of algal propagules tends to be widespread, the final zones are narrow and well defined, and this is to a great extent determined by interspecific competition. Since *Pelvetia canaliculata* can be killed at the top of its range by a few hot days in summer, however, physical factors certainly cannot be ignored.

It is also important to realize that communities are not static. Over a period of 10 years, Lewis (1977) found that numbers in the *Mytilus/Semibalanus/Patella* assemblage at the bottom of an exposed shore fluctuated violently. These fluctuations were due to such factors as irregular phases of predation by *Nucella lapillus* and the starfish *Asterias rubens*; irregular or seasonal effects of predation by birds eating limpets; and the blanket settlement of barnacle spat which smothered almost all organisms in some years but in others was hardly noticeable. Although these events were 'biological' factors, each was to some extent underlain by physical processes. Thus predation by *Nucella* and *Asterias* was limited in upshore extent by the relationship between their desiccation tolerance and the temperature regime in a particular year. Likewise, the clearance of weakened mussels and barnacles depended upon the timing and intensity of storm conditions. Over-concentration upon biological interactions is thus misleading.

The vertical distribution of any one species is in fact controlled by a complex interaction between biological and physical factors. An understanding of vertical zonation on the shore can therefore only be reached by seeking to understand the interplay between these two aspects.

Experiments to investigate vertical distribution

In many cases, investigations can best be organized in conjunction with an examination of the effects of wave action (see Chapter 3), so here we give some examples that can be applied to a variety of shores. Note that many of the experiments suggested in later chapters would be appropriate here.

Vertical distribution of the common species

To carry out a standard 'transect', use a cross-staff to establish appropriate vertical height intervals. At each height, random quadrats can be investigated for presence/absence, or abundance, of common species of algae and invertebrates. This information can then be used to plot kite diagrams, and to establish whether 'zones' are formed by single major species, or whether they contain distinct assemblages. Is there a three-zone structure?

Vertical distribution on a variety of shores

A comparison of the vertical height bands of some prominent species (e.g. algae, barnacles, limpets) on different shores gives some insight into controlling factors. Choose shores with, for example, different slopes, different aspects, different wave exposure or different rock type.

Vertical distribution of population structure

Many species show changes in size of individuals with vertical height. Juvenile limpets, for instance, are usually found low on the shore, while larger (females) are common higher up. Barnacles, on the other hand, show a more random distribution of sizes. Take one (or a few) species and investigate population structure in relation to vertical height. This approach will give some ideas about the factors that govern recruitment and the subsequent fate of individuals.

Behaviour of mobile gastropods in the field

If the shells of gastropods (e.g. topshells, winkles or dog whelks) are dried, they are easy to mark with quick-drying enamel so that individuals can be recognized later. Select a population from one level on the shore, transfer some upshore, some downshore, and replace some as controls. Then monitor vertical distribution of the three groups to see whether snails can move back to their normal zone.

Behaviour of mobile gastropods in the laboratory

The ready mobility of topshells and winkles under water in the laboratory provides opportunities for experiments on responses to various physical cues that may aid the snails in maintaining their vertical distribution. Given slopes of different angles, in what directions do snails move? Given light and shade as alternatives (simulating one facet of open rock v. crevices), do all species seek shade? Given either algae or inert alternatives such as plastic mesh, do all species exhibit thigmotactic (contact) responses?

3 Communities on the shore: the effects of wave exposure

Patterns of vertical zonation exist on most rocky shores, but the communities that make up the different zones differ radically from one shore to another. In north-west Europe, much of this variation in community composition seems to be related to the extent of exposure. On very sheltered shores, the rocks are almost totally obscured by macroalgae. On very exposed shores, however, the rock contours are more obvious because there are few plants but a dense cover of mussels and barnacles. To complicate matters, these differences in relation to exposure are not seen in other parts of the world. Another complication is that the three zones discussed in Chapter 2 show very great variations in vertical extent, and in some places the shore is not readily split into three zones at all. Again, much of this variation relates to exposure.

In this chapter we begin by describing some of the variation in community type that seems to be related to wave exposure; then we talk about possible explanations for the variation. We go on to discuss the communities found where rocky shores have special characteristics—periods of constant submersion (in pools), strong current flow, reduced light intensity and reduced salinity.

Communities in relation to wave exposure

A major problem, when dealing with any particular shore, is to know how exposed it is. One way in which biologists can estimate exposure is to use a 'biological exposure scale'. This consists of descriptions of communities which are typical of various categories of wave exposure. While such scales can be used very effectively as guides, it is important to note that they cannot be used to explain the effects of exposure. Any attempt to use them as an explanation results in a circular argument: a shore is exposed because it has a certain community pattern, and it has this pattern because it is exposed . . .

Among 28 sites round and near the Dale peninsula in south-west Wales, shores with similar patterns of communities can be related to categories of wave exposure on a scale from 1 (extremely exposed) to 8 (extremely sheltered) (Fig. 3.1). This scale, often called 'Ballantine's scale' after its inventor, has since been widely employed in north-west Europe, with suitable modifications in various situations, as an indicator of wave exposure. For example, it has been abbreviated to 5 points (Lewis, 1964), and it has been modified for use in western Norway.

Fig. 3.1 The variation in distribution of some common intertidal organisms in relation to exposure, as shown in south-west Britain. Note that the vertical zones have been compressed to a constant height. Their real extents are shown in Fig. 3.2. Note also that in intermediate exposures, organisms form mosaics, and even in extreme shelter there may be patches with barnacles, more typical of exposure. Abbreviations: *Fve, Fucus vesiculosus* var. *evesiculosus; Fs, Fucus serratus; He, Himanthalia elongata; Ls, Laminaria saccharina; Me, Mytilus edulis; Ms, Mastocarpus stellatus; P, Porphyra* spp.; *Pc, Pelvetia canaliculata.* (After Ballantine, 1961.)

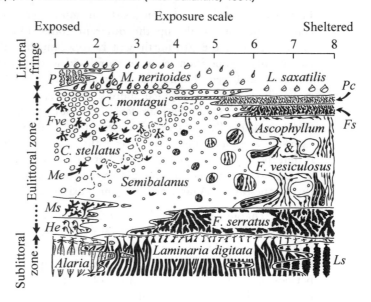

The composition of communities

The overall picture seen in north-west Europe with increasing exposure is one of gradual change from total cover by brown algae to dominance by barnacles, mussels and red algae (Fig. 3.1).

On sheltered shores (grades 7–8 on Ballantine's scale), fucoid and laminarian algae cover the shore so completely that no three-zone structure is visible, and there is no littoral fringe at the top of the shore. Underneath the thick algal cover, animals are abundant, but need searching for. Barnacles (*Semibalanus balanoides* and the Australasian species *Elminius modestus*) are moderately common in patches, accompanied in places by their predator the dog whelk, *Nucella lapillus.* Limpets are scarce and consist of relatively large individuals of *Patella vulgata.* Of the winkles, *Littorina mariae* is abundant on *Fucus serratus*, while *L. obtusata* is very abundant on *Ascophyllum.* The rough winkle, *Littorina saxatilis*, is common among the upper fucoids, while the large edible winkle, *L. littorea*, is common throughout the eulittoral.

This picture can be contrasted with that for semi-exposed shores (grades 4–5). These shores are similar to that at Plymouth, described in Chapter 2. In the sublittoral, kelps are still abundant, but consist now mostly of *Laminaria digitata*. In the eulittoral, algae are much scarcer, and the change in *Fucus vesiculosus* is particularly noticeable: instead of forming extensive cover, it is found in patches, often in a form that has no air bladders (forma *linearis* or *evesiculosus*). The barnacle *Semibalanus balanoides*, with its predatory dog whelks, is more widespread, so that there is a mosaic of fucoids, barnacles and bare rock. The upper levels of the eulittoral zone are dominated by two other barnacles, *Chthamalus montagui* (at the top) and *Chthamalus stellatus* (extending further down). Within the eulittoral, the limpet *Patella vulgata* is joined by two other species: *P. aspera*, often in tide-pools, and *P. depressa*. There is a substantial littoral fringe at the top of the shore, obvious because of the black lichen *Verrucaria maura*, but also containing a winkle not found on sheltered shores, the shiny black *Melarhaphe neritoides*.

On very exposed shores (grades 1–2), the dominance of brown algae is further reduced, except in the sublittoral where *Alaria esculenta* is characteristic. The red alga *Mastocarpus stellatus* replaces the zone of *Fucus serratus*, and turfs of other red algae, especially corallines, are abundant. The main part of the eulittoral is dominated by patches of barnacles and of small mussels, often with dense accumulations of *Nucella*, and abundant limpets. The barnacles are now primarily *Chthamalus*, of both species. Patches of *Fucus vesiculosus* still occur, but these are all very stunted, and all of the form *evesiculosus*. The winkles *Melarhaphe neritoides* and *Littorina saxatilis* are abundant, and spread from the eulittoral right up into the littoral fringe.

Although this change from total cover by brown algae in shelter to a mosaic structure in moderate exposure and lack of algae in great exposure is normal for northern Europe, the situation is very different in other parts of the world. On the opposite side of the Atlantic, in the Bay of Fundy, the biomass of the alga *Ascophyllum nodosum* does indeed decrease with increasing exposure; but even on exposed headlands it is still the dominant organism in the eulittoral zone (Thomas, 1994). The barnacle *Semibalanus balanoides* shows a slight increase in abundance from sheltered to exposed shores, but the genus *Chthamalus* is absent, and exposed shores are in no way as dominated by barnacles as they are in Europe.

Even these differences between the two sides of the Atlantic pale into insignificance when comparisons are made with other rocky shores in the temperate zone. On the Pacific coast of North America, communities show quite different relationships to wave exposure. Exposed coasts are, indeed, dominated by barnacles and mussels in the upper eulittoral, but often these may be joined by a striking brown alga, *Postelsia palmaeformis*, which forms 'groves' even in fierce surf. This alga, a laminarian, has erect stipes that bend in the breakers without themselves breaking. The lower eulittoral usually has a high degree of algal cover, but in southern parts of California even this is absent. On more sheltered shores, the upper eulittoral usually remains animal-dominated, though there may be large patches of a fucoid alga, *Fucus distichus*. On other sheltered shores, though, the sparse mussel and barnacle communities are surrounded by bare rock. The dense fucoid cover, so typical of sheltered shores in north-west Europe, is entirely absent.

In Australia, too, an equivalent to the blanket cover of fucoids in the north Atlantic is missing. Here the eulittoral zone on moderately exposed shores is dominated by animals such as barnacles, ascidians and tube-dwelling polychaetes. On sheltered shores there are growths of algae, but these are mostly encrusting forms such as *Hildenbrandia* and *Ralfsia*.

The vertical range of communities

The changes in community composition that occur with changes in exposure are accompanied by striking changes in the vertical levels of zones on the shore. In north-west Europe, all the zones become greater in vertical extent as wave exposure increases, and thus are found at greater heights above chart datum (Fig. 3.2). The upper limit of laminarians (the top of the subtidal zone) rises. The upper limit of barnacles (the top of the eulittoral zone) rises substantially. The upper

Fig. 3.2   Variations in the height of zones on the shore in relation to exposure. (After Lewis, 1964.)

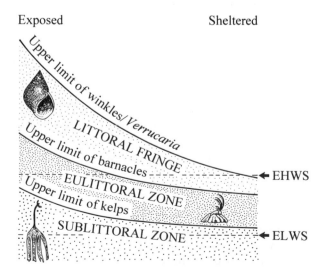

limit of *Littorina* and *Verrucaria* (the top of the littoral fringe) may rise dramatically, so the vertical extent of the shore may increase from a few metres in shelter to 30 m or more in extreme exposure. Once again, however, comparisons with the Bay of Fundy show interesting differences. Here, the tops of the eulittoral zone and the littoral fringe do indeed rise with increasing exposure, but the top of the sublittoral zone appears to stay at a constant level. The vertical ranges of many individual species also show no change with increasing exposure, and some of these include dominant organisms such as *Ascophyllum nodosum*, *Fucus vesiculosus* and *Littorina obtusata*. Overall, zonation patterns tend to be essentially constant with respect to exposure.

These differences in the response of zonation patterns to different degrees of wave action suggest that the distributions of fauna and flora are far from being solely under the direct control of physical forces. We discuss possible causes in the next section.

Explanations of the effects of wave exposure

The importance of wave forces

There is no doubt that the physical differences between exposed and sheltered shores directly influence the distribution of some organisms. Delicate algae typical of, say, gently sloping sheltered shores cannot survive the battering that occurs on exposed headlands. Some of the gastropods common in dense algal beds cannot adhere to the substrate when water movements are very rapid (see p. 42). Furthermore, the characteristics of the water found in sheltered inlets are often very different from those of water passing projecting cliffs. In enclosed bays, water will be warmer in summer, more turbid and often less saline and less oxygenated than offshore water. Inshore turbidity often leads to deposition of sediment on the rock surface, so that the physical habitat presented to a colonizing propagule here is quite different from that on a wave-exposed outer coast.

In response to some or all of these differences in physical environment, as well as to differences in biological interactions, the organisms that are able to tolerate both exposed and sheltered conditions show considerable ranges of form and size. For example, individuals of the dog whelk, *Nucella lapillus*, collected from exposed headlands are hardly recognizable as the same species found in shelter: they are small and squat, compared to the long shells typical of shelter. The size of the aperture is relatively much bigger on exposed shores, indicating that snails have a large foot, allowing better adhesion (p. 142). The changes in the alga *Fucus vesiculosus* have already been mentioned when describing exposure scales. In shelter it has long fronds with many air bladders which help the fronds to float when the tide is high. On exposed shores the plants are stunted and do not possess air bladders. Laminarian algae, also, have different morphologies in different environments. *Laminaria hyperborea*, for example, has a blade divided into 'fingers' in exposure, but when found in shelter it grows a rounded, entire blade (p. 67).

Can these variations in form and size be explained by physical constraints, or are they due to biological effects? Some of them have now been investigated in sufficient detail to allow judgement to be made of the importance of the physical effects of wave exposure in their determination. We begin by discussing the limits to overall size, and then go on to discuss changes in morphology.

On exposed shores, the rapid movement of water in pounding waves might be expected to cause problems for larger organisms. But in fact, the drag forces caused by fast rates of flow do not impose limits on size. Drag *does* increase with increasing size, but since the increased force is spread over a larger area, force per unit area is unchanged in larger organisms. Much of the force experienced in waves is actually due to the acceleration of the water. The accelerational force increases as length increases until, if length is too great, force per unit area becomes too great, and the organism will break or dislodge.

These points were shown well in a study of animals on the Pacific coast of North America (Denny *et al.*, 1985). After measuring the forces due to acceleration and to drag in a steady rate of flow, it was possible to calculate the probability of dislodgement for a variety of species. For limpets, sea urchins and mussels, the risk of destruction increased substantially for larger individuals: the sizes of these animals found on the shore were to a great extent determined directly by mechanical constraints imposed by accelerational forces. For several species of snails, however, the situation was very different. Their ability to cling to the rocks was very limited, and even moderate velocities and accelerations dislodged small snails as well as big ones. In their case, the waves did not therefore determine size, but they did affect overall distribution, since the snails could not adhere in very exposed conditions. How could the snails exist at all on these exposed shores? This is explained quite simply because the snails withdrew to sheltered microhabitats such as crevices when conditions were rough.

Another study on the Pacific coast of North America examined the possibility that the overall size of smaller algae might be limited by hydrodynamic forces. To measure the breaking force of each species, Gaylord *et al.* (1994) attached a spring balance to algae in the field, pulling until the plants broke, and then read off the force applied. In the laboratory, they measured drag forces on the algae under various conditions in a flow tank. Even at exposed sites, the calculated risk of breakage from drag forces alone was slight. However, because the algae had large inertia coefficients, they experienced large accelerational forces in exposure, and these, not drag forces, were calculated to form the major influence on maximum size (Fig. 3.3).

Fig. 3.3   The calculated probability of the alga *Iridaea flaccida* surviving for 3 months where wave height averages 2 m. The dashed line (probability calculated from drag alone) suggests almost unlimited survival, but the solid line (which also includes accelerational force) shows that many of the larger plants would in fact be torn off. (After Gaylord *et al.*, 1994.)

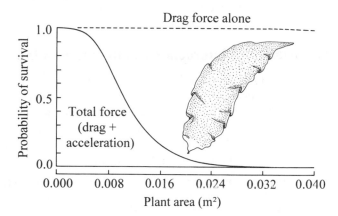

In spite of the large accelerational forces produced by waves, increase in size can be achieved even in very exposed habitats, by a change in form. Particularly when body shape is flexible, drag and accelerational forces can be minimized if, for instance, the organism is aligned with its length parallel to the flow. A study of the laminarian *Nereocystis luetkeana* nicely defined the functional consequences of blade structure in varying water currents (Koehl and Alberte, 1988). This giant kelp, common on the Pacific coast of North America, has long blades which arise from a float at the end of a long thin stipe. In sheltered areas the blades are wide and undulate, while in exposed areas they are narrow and flat. The behaviour of the blades in a current was observed by towing plants behind a boat: the narrow blades clumped together into bundles that were narrower than those formed by the wide blades, and they showed little tendency to flap. The undulations on the wide blades promoted pronounced flapping, which aids such factors as gas exchange. However, since flapping increases drag, these wide blades experienced more drag than the narrow ones. Koehl and Alberte calculated that if the algae at current-swept sites had possessed wide undulate blades, drag forces would have been sufficient to break the stipes.

In Chile, two common kelps show quite different distributions in relation to differing wave exposure (Santelices, 1990*a*). *Lessonia nigrescens*, a species that has multiple stipes arising from one holdfast, is dominant in very exposed conditions. The normal form of *Durvillaea antarctica*, which has only one or a few stipes, is not so common in severe exposure. This species also occurs as a morph with multiple stipes, however, and this morph co-exists with *Lessonia* at exposed habitats. The two species of kelp probably compete for space, and multiple-stemmed forms are better able to compete in the face of extreme wave action because less weight is supported on each stipe. Competitive displacement, mediated by wave action, could therefore explain their distributions.

From these examples, it is evident that the physical effects of waves cannot be considered in isolation, though there are situations in which they are overriding. They must be integrated with the effects of biological interactions, and we now turn to these.

The importance of interactions between species

In the north-east Atlantic, one of the major causes of the change from high fucoid cover on sheltered shores to dominance by barnacles, limpets and mussels on exposed shores, is the increase in effectiveness of limpets as grazers along a gradient of increasing exposure (Fig. 3.4; Hawkins *et al.*, 1992). On sheltered shores, there is only a low density of *Patella vulgata*, although the individuals are large. On exposed shores there are

Fig. 3.4   Hypothetical cycles showing relationships between patellid limpets and fucoid algae on British shores. On exposed shores, the cycles favour limpets, and fucoids are scarce. On moderately exposed shores, patches of fucoid algae promote limpet settlement and growth, but in turn the limpets may graze away the algae and prevent their resettlement. On sheltered shores, the cycles favour the algae and only occasional patches are grazed. Stippled areas show the distribution of fucoid algae. (After Southward and Southward, 1978; Hawkins and Hartnoll, 1983.)

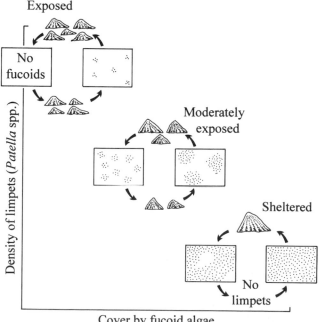

three species of *Patella*, and although individuals are often small, population density is high. The effectiveness of *Patella* in keeping particular zones clear of fucoids has already been mentioned (p. 26) and limpet grazing is discussed further on p. 81. Yet the question remains: why do limpets fail to control fucoid settlement and growth on sheltered shores? Generally, the population structure of limpets on sheltered shores suggests that individuals, once settled there, are long lived, but that settlement does not happen very often. It may be that the fucoids themelves deter settlement by the sweeping action of their fronds, or the silt trapped underneath them may prevent the larvae from attaching. Alternatively, it may be that few larvae actually reach really sheltered bays—this problem, under the title of 'supply-side ecology' is discussed on p. 177. Whatever the limiting factors, they are of extreme importance, and require investigation.

In the north-west Atlantic, where fucoid seaweeds are often abundant even on wave-exposed coasts, the biological interactions that control community structure are quite different (Chapman and Johnson, 1990). One of the major reasons for the difference from European shores is the lack of patellid limpets. The major grazers are winkles, *Littorina* spp., and they feed mainly on ephemeral algae. The barnacle and mussel populations, although significant, are reduced by predation from the dog whelk, *Nucella lapillus*. Both grazing and predation maintain free space on the rock surface, and allow settlement and growth of fucoid algae even on exposed shores. This straightforward explanation needs, however, to be tempered by consideration of the distribution of the alga *Ascophyllum nodosum*. This is common in exposure as well as in shelter, thus showing quite a different pattern to the one in Europe. Yet after experiments in the Bay of Fundy in which *Ascophyllum* has been removed, there has been no recolonization in 10 years. Sporelings that settled were rapidly grazed by *Littorina* spp. and other gastropods, so while it is possible that *Ascophyllum* will eventually reappear, this must be a very long-term event (Thomas, 1994).

On the Pacific coast of North America, algal community structure appears to be controlled by a hierarchy of competitive vigour (Dayton, 1975). In exposed areas, the kelp *Lessoniopsis littoralis* outcompetes other species: when it is experimentally removed, the rock is colonized by ephemeral species and by another brown alga, *Hedophyllum sessile*. In areas of moderate exposure, *Hedophyllum* outcompetes the ephemeral algae, and its growth is vigorous enough to overcome the effects of grazing molluscs: removal of the chiton *Katharina tunicata* has no detectable effect upon the regrowth of *Hedophyllum* in experimentally cleared plots. The series of competitive changes from shelter to exposure is reminiscent of the series from high shore to low shore in fucoid beds in north-west Europe (p. 24).

Conclusions: why are communities on exposed and sheltered shores different?

In many of the cases considered above, individual factors have been isolated and shown to be important. But in seeking for general answers, the major problem so far has been the small scale on which experiments have been carried out. Investigations have, inevitably, been carried out at individual sites, and conclusions drawn from results obtained in relatively limited areas. Within these areas, there seems little doubt that wave action can have a dual effect. It influences structure and community composition directly, by eliminating organisms that cannot withstand the large accelerational forces found on exposed shores. It also has indirect influences, changing the balance of competition, predation and grazing pressures. If this is so, we particularly need to know how representative are the areas that have been investigated. On the Pacific coast of North America, a survey of 20 sites over a range of more than 800 km showed that there was enormous variation in features such as total percentage cover by organisms, abundance of individual species, and vertical distribution of species, and that these could not always be related simply to degree of wave exposure (Foster, 1990). It is difficult to absorb such overwhelming variation into any general model of explanation based upon interactions between a few major species, and we can conclude that, as yet, claims for generality in understanding how ecosystems function are not appropriate. In other words, the models discussed above are probably adequate for the immediate areas in which they were studied. Whether they are appropriate for wider areas such as 'the west coast of North America', or 'the shores of north-west Europe', has yet to be shown. We shall consider various effects upon community structure in more detail in Chapter 8.

The communities of special habitats

Although wave exposure has been shown to have very important influences (both direct and indirect) on the distribution of fauna and flora, many other physical characteristics of a shore govern the distribution of communities upon it. Rock type and texture determine how well water drains from the shore, and affect the microhabitats in which propagules can settle and grow. Conditions during submersion are to a great extent decided by the position of the shore with respect to water currents: even on subtropical shores, cold-water currents can deter many species (including *Homo sapiens*). Conditions during emersion are of course affected by climate, and hence by latitude, but also by the direction in which the shore faces (the aspect). A northerly aspect (in the northern hemisphere) on a shore with high cliffs, means that the sun will seldom cause much desiccation, while on a southerly aspect the cliffs may trap the heat so that drying and overheating may become limiting.

Even the time at which low water occurs during the spring-tide period may be a critical factor. The times at which this, and other features of the tides, occur is relatively constant for any one locality (see p. 5). Where low water of spring tides occurs in the middle of the day, evaporation and desiccation may be intense; but in places where it occurs in early morning or late afternoon, there will be less physical stress.

Besides these factors that have an overall influence on the shore, there are many more localized features that affect the microclimates in which organisms live. The discussions so far have really treated the shore as if it were a regularly sloping expanse. But there are very few shores like this—most have crevices, gullies, overhangs and maybe caves, pools, damp areas and areas of freshwater drainage. Each of these features may considerably alter the conditions from those on the open rock face. We take four examples, to show the degree to which individual variables can assume an overriding importance. Examination of these special habitats can also provide insight into the factors governing distribution in more 'normal' habitats on the shore.

Rapids and the effects of current speed

Although some of the effects of wave action on exposed shores have been attributed to velocity of water motion, we have seen that many consequences derive from accelerational forces (p. 36). It is difficult to disentangle these effects on normal shores, but at restricted sites the direct effects of prolonged high velocity can be investigated. There are many 'rapids' on the west coast of Scotland and in Norway, often with currents that reverse on rising and falling tides. In the south-west of Ireland, the rapids at Lough Hyne have been investigated in detail (Kitching, 1987).

The rapids at Lough Hyne occur because the channel that connects the lough to the sea is constricted enough to introduce a time delay in the equalization of water levels on both sides of it. As the tide falls outside the lough, water pours out through the rapids, reaching a maximum current speed of 3 m s^{-1}. When the tide rises again, it first causes a momentary slack water in the rapids, and water then pours into the lough, this time reaching a maximum current of 2 m s^{-1}. Current speeds decrease in velocity both up and down the rapids channel from a point known as the Sill, causing differential distribution of sediment and effects on water temperature, so that even here there are complicating factors to add to the direct effects of water velocity. But animal and plant species show very clear distributions in relation to current speed at Lough Hyne (Fig. 3.5), and consideration of three of these gives some impression of the causes of distributions and the problems in determining them.

The hydroid *Sertularia operculata* grows attached to boulders in strong current on the Sill, but is scarce in places where the current is less strong and disappears altogether just inside the quieter waters of the lough itself. When boulders with *S. operculata* were moved into the lough, the hydroids became clogged with fine sediment within a week and most of the colonies were dead within 9 weeks. Diatoms grew over the outside of the hydroids, and other plants and animals invaded the colonies. Thus although the direct cause of death was not certain, the settlement of fine particles probably played a major part.

Fig. 3.5 The distribution of laminarian algae in relation to current velocity in the 'rapids' at Lough Hyne, Ireland. The bottom diagram shows the currents at the time of low tide in the sea (to the right), when water is flowing out of the lough at maximum velocity. *Saccorhiza polyschides* is widely distributed where current speeds are moderately high, but it does not penetrate into the calm waters of the lough or the creek that connects to the sea. In the fastest water, particularly where water breaks over a 'sill' (shown by stipple), *S. polyschides* is replaced by *Laminaria digitata*. In calm water it is replaced by *Laminaria saccharina*. Current velocity is shown as m s^{-1}, and 'S' indicates a very slight current (< 0.04 m s^{-1}). Arrows show the direction of flow. The solid line shows the high water mark of spring tides; the dashed line shows low water mark of spring tides. (After Bassindale *et al.*, 1948.)

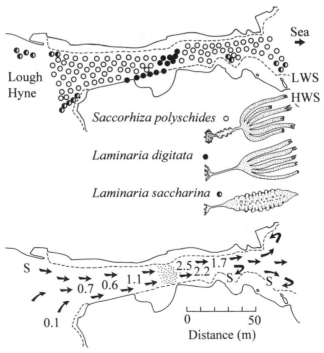

In the case of the topshell *Gibbula cineraria*, found on the blades of *Saccorhiza polyschides*, there is strong evidence that its complete absence in strong current is simply because it cannot hold on. In the rapids area, it is confined to stations around the corner from the main channel, at each end. When snails from these regions were allowed to settle on plants of *S. polyschides* in quiet water, and were then transferred to stations within and outside the rapids, the difference in numbers still clinging to the alga was dramatic. In the rapids stations, almost all the snails disappeared in 4 h, while at sheltered sites most remained.

The sea anemone *Corynactis viridis* is common at intermediate current speeds in the rapids area. Boulders with many anemones on them were moved to a sheltered site just inside the lough, in a variety of positions, and examined over a period of 2 years. Those placed either on the sea floor or on a raised platform, facing upwards, regressed and died. Those facing downwards on a raised platform remained large and active. It seems likely that as with the hydroid *Sertularia*, sediment was involved in causing deaths, but the growth of microalgae may have been important. The effects of excessive illumination may also have played a part, since anemones placed facing upwards in floating cages in the rapids remained shut and refused to feed.

Tide-pools and the effects of constant submersion

Because the communities of tide-pools remain constantly submerged, the pools have sometimes been regarded as natural aquaria, reflecting life in the sublittoral zone. But pools high on the shore undergo the consequences of prolonged separation from the main body of the sea (Huggett and Griffiths, 1986), and conditions within them fluctuate dramatically. On the Atlantic coast of the Cape Peninsula, South Africa, the maximum temperature during the day increases from about 16°C in pools near MLWS, to 30°C near MHWS. Maximum temperature occurs during mid-afternoon because the water continues to absorb heat from the surrounding rocks even after air temperature starts to drop.

The situation with oxygen concentration, carbon dioxide concentration and pH is less simple because here the biota directly affect the variables as well as being affected by them (Fig. 3.6). During the day, algae produce oxygen by photosynthesis, and oxygen concentration may rise to three times its saturation value, so that it is actually released: bubbles can often be seen rising from dense beds of algae. The algae also absorb carbon dioxide, so that carbon dioxide concentrations fall (Morris and Taylor, 1983). One of the effects of this fall in carbon dioxide concentration is a rise in pH, which can be extreme. Values greater than 9 were recorded on the Isle of Cumbrae, Scotland, and pools in the Baltic have shown values greater than 10. At night, when photosynthesis stops, pools can show equally drastic changes in the opposite direction. Respiration absorbs much of the available oxygen, and values of 1–5% saturation have been recorded. Carbon dioxide levels increase, and pH may fall below 7. These changes are not as simply related to height on the shore as those in temperature. In 33 pools on the Cape Peninsula, maximum oxygen saturation occurred near MLWN. Pools both above and below this showed smaller effects of photosynthesis, because the higher ones contained few algae while the lower ones were emersed for only very short periods.

Fig. 3.6   Conditions in a high-shore tide-pool on Great Cumbrae Island, Scotland, during July, compared with those in the sea. The pool is dominated by the algae *Cladophora* (shown), *Corallina* and *Chaetomorpha*. Photosynthesis during the day produces high oxygen levels, lowers carbon dioxide and thus raises pH. Respiration at night reverses these changes. Solid circles show values in the pool; open circles show values in the sea. (After Morris and Taylor, 1983.)

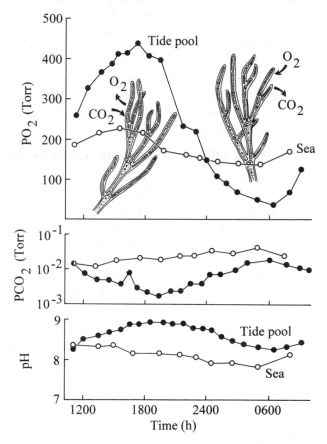

There are often other extremes to which the inhabitants of tide-pools are subjected. In the long term, salinity may increase from evaporation, or decrease from rainfall and water seepage, and these changes can be overwhelming in high-shore pools. Vertical gradients of many variables can arise within the pools, so that different depths within one pool can experience quite different conditions. Thus, although tide-pools offer protection from desiccation, they also present their inhabitants with serious challenges. Not surprisingly, the fauna and flora of tide-pools vary enormously, particularly in relation to height on the shore, but also in relation to shelter, shading, size and depth.

A great deal of the variation in the biota that occurs between different tide-pools is determined by the time for which they are isolated from the sea. Figure 3.7 shows four pools in south-west Ireland (Goss-Custard *et al.*, 1979). Here the pools above MHWS are densely covered by the green alga *Enteromorpha intestinalis*, and some have large populations of the copepod *Tigriopus fulvus*—densities as high as $720 \times 10^3/m^2$. There are few other invertebrate species in these high pools. In pools between MHWS and MHWN, encrusting coralline algae and the branching species *Corallina officinalis* cover most of the rock surface. Limpets are common, and are effective grazers. Experimental removal of limpets from these pools resulted in a dense growth of *Enteromorpha* over a period of 2 months, so that the pools came to resemble those higher up. Pools between MHWN and MLWN contain *Laminaria* spp. as well as other large brown algae, and *Corallina* forms an extensive 'undergrowth'. Limpets are scarce here, but there is a wide variety of small invertebrates, especially those inhabiting the *Corallina*. Below MLWN, the algal flora is further enriched, especially by the addition of red species. Here the community is very similar to that found sublittorally, but conditions are evidently particularly favourable for some species of animals and plants. The blue-rayed limpet, *Helcion pellucidum*, for instance, is much more common here on *Laminaria hyperborea* than in the sublittoral. These low-level pools could, indeed, be regarded as 'sublittoral aquaria'.

Fig. 3.7   A diagrammatic view of four tide-pools on the exposed headland of Carrigathorna, south west Ireland. (After Goss-Custard *et al.*, 1979.)

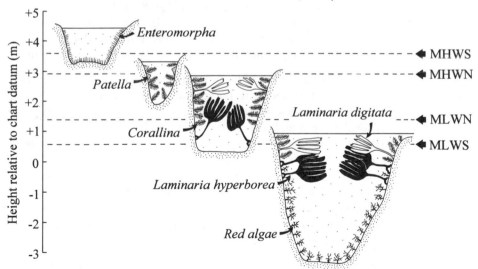

The rigorous conditions in the higher level pools probably prevent many species from colonizing. Experiments on the copepod *Tigriopus fulvus*, for instance, suggested that salinities and temperatures can be critical. In the laboratory, this species was found to survive for 15 days in salinities ranging from 42 to 90, but died after 84 h in distilled water, and sank to the bottom in salinities greater than 90 (see p.47 for explanation of units of salinity). It could recover after some hours if returned to sea water (salinity 35) after this. In tests with a slowly rising temperature, the death point was 32°C at a salinity of 34, but this rose to 41.8°C at a salinity of 90. Thus high salinity helps to enable *Tigriopus* to withstand high temperature, as is appropriate for an animal living at a level where insolation and evaporation may be considerable.

The physical conditions within any one pool will vary over an annual cycle, but will also be subject to random 'disturbance' events (Dethier, 1984). For instance, excessive wave action, overheating, battering by current-borne logs or boulders, or invasions by predators or grazers can significantly disturb the normal communities. Just as on open coasts considered in more detail later (p. 164), this disturbance may mean that individual opportunist species quickly come to dominate individual pools. If the disturbance rate is high, and rates of recovery are low, these events will make it impossible for any one species to occupy all the pools at any one time. Part of the amazing diversity of tide-pool systems may therefore depend upon a dynamic state structured by disturbance.

Sea caves and the effects of reduced illumination

On normal shores, light intensity decreases with depth, but because this decrease is also associated with changes in spectral composition, it is very difficult to know whether the changes in algal flora that occur with depth are determined by the reduction in light or by changes in its quality (see p. 21). Caves provide situations where light intensity changes without changing in wavelength, and therefore provide useful natural experiments.

The most obvious feature of biota inside sea caves is a rapid decrease in numbers of algal species with horizontal distance from the cave mouth. In a tidal cave in south-west Ireland, fucoids and laminarians persist only for the first 6 m, but even there the fucoids are so stunted that it is impossible to distinguish the species (Norton *et al.*, 1971). At 8 m inside, fucoids and laminarians are absent, but the green alga *Ulva lactuca* and some red algae such as *Plumaria elegans* and *Lomentaria articulata* are present. At 20 m inside, there is still a variety of red species, including *Plumaria elegans* and the encrusting *Hildenbrandia prototypus*, as well as the green *Cladophora* sp. Some of these penetrate to 26 m, but no algae at all reach 40 m from the mouth.

Is this order of penetration determined by the gradual reduction of total illumination? For many species this is too simple an explanation, because the order of penetration into the cave is not the same as that found outside the cave with increasing depth. Red algae, for instance, penetrate into lower light intensities in the cave than they do with depth. There is good evidence that total illumination is the critical factor for laminarians, however: *Laminaria hyperborea* penetrated into the cave to a distance where overall illumination was 1–4% of that outside, and this coincided very closely with conditions at the lowest depth limit of the species on the open coast (17 m), where illumination was estimated to be 2–6% of that at the surface.

The fauna inside the cave is less restricted than the flora. Barnacles, anemones, tubeworms and starfish are found as far as 80 m inside, and many species penetrate to 40 m. Many of these, as well as several algae, extend higher up the side of the cave with increasing distance into it. Presumably this reflects the lack of desiccation, but this is a subject that needs further investigation.

Estuarine shores and the effects of reduced salinity

On typical marine shores, most organisms experience 'normal' salinity most of the time. 'Normal' sea water contains approximately 35 g of salts in 1 kg of water. This used to be expressed as '35 parts per thousand', or 35‰, but now that salinity is measured from conductivity, it is expressed without units, simply as a salinity of, say, 35.0. High on the shore, salinity in tide-pools may rise by evaporation, as discussed on p. 44, or it may fall where streams run over the shore. For the majority of the shore, however, these effects are minor, and to see the gross effects of changes in salinity, we need to view the situation in estuaries.

One of the major characteristics of estuaries is a reduction in numbers of species with distance from the sea. A good example is provided by the Severn Estuary, Britain (Bassindale, 1943). Here, the number of marine species recorded on rocky shores diminishes from about 400 outside the estuary to only one at the head. This type of reduction has classically been attributed to a salinity effect: average salinity values decrease, and salinity variation increases, with distance upstream, and many laboratory experiments have shown that both these factors have deleterious effects on marine animals and plants. For example, the larvae of the scallop (*Pecten maximus*) can tolerate gradual decreases in salinity but abrupt changes cause high mortalities (Davenport et al., 1975).

However, many factors must be taken into account when explaining distributions recorded from an estuary. Salinity reduction is but one of the characteristics of estuaries. Turbidity often rises (spectacularly so in the Severn Estuary, where average values of suspended sediment reach 2 g l^{-1}) and temperature, current speed and substrate type usually also change radically. Turbidity, in particular, may have many effects. There is usually a zone where turbidity is maximum, somewhere in the mid-reaches, although it may move up and down with tidal flow. In this zone the plankton is severely reduced, so that food for suspension feeders is sparse and filtering mechanisms tend to become blocked with silt. To attribute the reduction in species numbers purely to salinity is therefore to confuse the situation by attempting to oversimplify it.

This point is reinforced when various estuaries are compared. For example, the alga *Fucus ceranoides* is found much further upstream than *Fucus vesiculosus* in the Dee Estuary, Scotland, but in the Tay Estuary, and in England the Tees Estuary, *Fucus vesiculosus* is the one found further upstream. These differences do not seem to be related to any obvious differences in salinity regime. In the Baltic, however, salinity may well determine which species penetrate the furthest. Here, marine species penetrate into lower salinities than they do in estuaries, because the fluctuations in salinity are small, compared to those in estuaries.

Marine species that penetrate into estuaries often show considerable changes in their morphology and in their distribution in comparison with populations in the sea (Barnes, 1984). For example, estuarine individuals are often small. The mussel *Mytilus edulis* reaches a length of 110 mm near the mouth of the Baltic, but at the head, mature specimens grow no bigger than 20 mm. Among the algae, *Laminaria saccharina* reaches normal lengths of 1 m or more in sea water, but is only a few cm high at its limit of penetration into the Baltic. The cause of this dwarfism is not known, and it should be emphasized that some species show no such trend.

Vertical distributions of species in estuaries can be radically different from the typical picture described earlier in the chapter. In the Baltic, *Fucus vesiculosus* is the only fucoid present, and its zonation is quite different from that in the rest of north-west Europe (Voipio, 1981). Its upper limit lies at the top of the sublittoral zone, and from here it extends downwards to a depth as much as 10 m below water level. The reasons for this distribution are not clear, but the upper limit is probably governed by ice scour during winter.

Vertical zonation patterns in the Severn Estuary, Britain, also show great differences from 'normal' patterns (Mettam, 1994). To seawards of the estuary, *Laminaria digitata* occupies the top of the sublittoral zone, and at low water of spring tides parts of the kelp beds are emersed. Within the estuary, kelps are absent, and are not replaced by other algae so that the bottom of the shore below MLWS is bare except for encrusting barnacles. Another striking feature of many of the shores in the Severn Estuary is the predominance of the alga *Fucus serratus* on the mid shore (Fig. 3.8). On 'normal' marine shores this species is restricted to the lower part of the shore, and certainly below MTL. In some places in the Severn, it reaches up to MHWN, and this is not due to a general rise in vertical limits caused by exposure to wave action, since the shores in question are relatively sheltered. The dog whelk, *Nucella lapillus*, in contrast, is restricted to lower regions of the shore at upstream sites: at its upstream limit, it is found only below MLWN, whereas outside the estuary it is common up to MTL.

The factors that govern these distribution are likely to be connected to salinity only indirectly, because almost all estuarine organisms (unlike marine ones) show a remarkable tolerance of low salinity and salinity variation. Major factors may include food supply, competition, predation, tolerance of silt and many others. Because estuaries have limited numbers of species and 'odd' distribution patterns, the study of estuarine organisms on rocky shores may allow experimental testing of some of the ideas about factors that govern distribution on marine shores, where ideas are hard to test because of the high species diversity.

Fig. 3.8 The relative abundance of *Fucus serratus* at different tidal levels in the sea and in the Severn Estuary, Britain. The transect sites begin in West Wales (the sea) and end at Gloucester (the head of the estuary), but are not spaced at equal distances. Kite diagrams show abundance on a five-point scale, the maximum point allowing the kites to touch. Solid kites show the north shore of the estuary; dashed kites show upper limits on the south shore where these are higher. Tidal levels are shown relative to OD. Note that in the upper estuary, MLWN (dashed) is *lower* than MLWS. *Fucus serratus* is more common above OD within the estuary than in the sea. (After Mettam, 1994.)

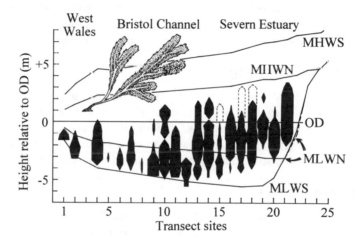

The investigation of communities

The communities available for study will vary enormously depending upon locality. Here we suggest some topics that it should be possible to investigate on a wide variety of shores.

Comparison of exposed and sheltered shores

The communities on opposite sides of a headland will usually provide a striking contrast in the effects of wave exposure. If sufficient time is available, transects on two shores will allow detailed comparison. Otherwise, community composition in quadrats at MTL will allow contrasts to be assessed. If a graded sequence of exposure is available, this will allow examination of changes in zonation of organisms, as well as their replacement at opposite ends of the scale.

Vertical range of communities

Without examining community composition, examination of the vertical distribution of dominant species at a range of different exposures gives a good impression of the effects of wave action. Compare the vertical ranges, in relation to chart datum, of *Laminaria* spp. (sublittoral), barnacles (eulittoral) and the indicators of the littoral fringe—the lichen *Verrucaria* and the winkles *Littorina saxatilis* and *Melarhaphe neritoides*—at as many sites as possible. Use exposure scales to assess wave exposure.

Tide-pool communities

If pools are available at a variety of heights, select a range from well into the littoral fringe down to the low shore. Examine the biota within each one, then attempt to see how fauna and flora correlate with physical variables: shape, size, aspect, depth, volume, tidal height. If there is time and equipment, analysis of chemical changes within a variety of pools is instructive; measure oxygen, pH and temperature.

Caves and overhangs

Very few shores have useful caves, but the effect of overhangs in diminishing desiccation can usually be investigated. Compare communities on steep faces of various aspects, and then make comparisons with regions that are really protected from sunlight by overhangs. Are communities different in composition; or denser; or at a different height?

Estuarine shores

Few leaders of field courses would choose to limit their students to estuarine rocky shores. But examination of at least one estuarine shore gives a new perspective to the ecology of more 'normal' shores. Make a comparison with a marine shore of the same type of wave exposure. Are there any animals and plants not found in the marine shore? Does zonation differ? What are conditions like in terms of deposited silt, salinity, tidal currents? Are the communities more or less species-rich?

4 Algae, the primary energy sources

The most obvious components of rocky-shore communities are either macroalgae or encrusting animals such as barnacles. On some sheltered shores, particularly in north-west Europe, the algae dominate totally, at least in terms of biomass. Although this dominance is not found in all other parts of the world, nor is it seen on wave-exposed shores, it emphasizes the point that algal growth is a major factor influencing shore ecology. There are littoral species of lichens, mosses and angiosperms, but while they are often abundant, they are not very significant in terms of food supply. Algae provide a primary energy source, and thus form much of the basis for intertidal food webs.

To some extent, though, the great biomass of the brown, green and red algae—the macrophytes—gives a misleading idea of their importance, if we think in terms of energy supply. Shores which do not have very abundant macrophytes usually have a film of 'microalgae'—a collective term that includes the blue-green algae, diatoms, other protists such as the euglenoids, and the spores and sporelings of macroalgae. As we shall see, these microalgae are often more important than the macrophytes in providing a food supply for the grazers because, in spite of their low biomass, they have very high rates of production. We begin this chapter by discussing the macrophytes, but come back to the microalgae at the end.

The distribution of littoral macroalgae

Rocky shores show many features in common in terms of plant distribution, regardless of continent. However, the distribution of some genera is limited (Fig. 4.1), and this leads to a few important differences round the world which we need to discuss here.

On most exposed shores, the shore appears relatively bare of macrophytes, although microalgae may be abundant. Only at the bottom of the littoral zone is there a rich assemblage. Here and in the sublittoral are the brown kelps, usually Laminariales such as *Laminaria*, *Alaria*, *Ecklonia*, *Nereocystis* or *Macrocystis*, often with a belt of short red algal turf above them. *Laminaria* is dominant in Europe and north-east America, but on the Pacific coast of America it is joined by several genera. *Sargassum, Egregia* and *Lessoniopsis* are typical large forms. *Postelsia* is exceptional in that it is the only laminarian to colonize the mid shore, where its upright stipes look like small versions of palm trees in the battering waves. The giant *Macrocystis*, which may grow to a length of 20 m, forms extensive offshore beds, famous for being the playground of sea otters. Towards the southern tip of South America, the laminarians are accompanied, and to some extent replaced, by the genus *Durvillaea*, which is also a brown alga, belonging to the Fucales. *Durvillaea* may grow to 10 m, has large blades, and is anchored by a massive holdfast. This genus is Antarctic in distribution, and is common in Australia, New Zealand and southern Africa. These southern areas also have further laminarian genera such as *Ecklonia*.

Fig. 4.1   The occurrence of some of the major genera of brown algae.

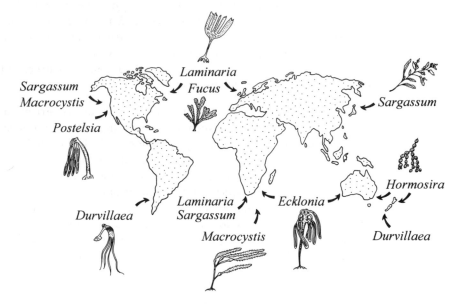

On more sheltered shores, the diversity of algae increases, and the whole shore may show extensive algal cover. In many cases this is due to growth of brown algae of the Fucales. In Europe and north-east America, *Ascophyllum* and *Fucus* are dominant, and 100% cover is often found in sheltered inlets. On the Pacific coast of North America, *Fucus* is still found, but only in restricted zones, and much of the shore is dominated by animals such as mussels, barnacles and oysters, and by turfs of red algae. The situation in southern Africa is very similar. The sheltered shores of Australia and New Zealand have their own characteristic species, in particular *Hormosira banksii*, which, although a fucoid, has the appearance of a necklace of green spheres—in fact its common name is Venus's necklace. The lower parts of the shore in

Australia, southern Africa and South America tend to be dominated by sea squirts such as *Pyura* rather than by algae.

On all shores, the variety of species increases towards the sublittoral, and—apart from the kelps—red algae tend to take over from the browns at the bottom of the shore. Green algae do not usually occur in large masses, but for all that some of their genera are world-wide: *Enteromorpha*, *Ulva* and *Cladophora* all occur in Europe, America, Africa and Australasia.

How algae fit into littoral food webs

Primary production by the algae of the littoral zone can be substantial. What happens to all this production? Grazers eat a large fraction of the microalgae, but they consume a decreasing amount of the production as plant size increases (Fenchel and Jørgensen, 1977). Instead, most of the biomass of large plants is degraded to detritus, and enters the detritus food chain (p. 56). Detritus, and the presence of detritivores, are therefore major factors to be examined in intertidal ecology. Why grazers do not consume high proportions of macrophytes is also a crucial question in terms of algal success, and we discuss this point first.

Why do grazers seldom eat algae? Algal defences

As we will discuss on p. 81, grazers can have significant effects on macrophyte distribution on rocky shores. Yet they seldom consume adult algae in large quantities. There are exceptions, such as the winkle *Littorina obtusata*, but the general distinction from the effects of grazers on microalgae is striking. How do algae survive in the presence of herbivores? There are only three basic ways in which they can do this: they can escape by using refuges either in space or time; they can in some way 'tolerate' the herbivores; or they can deter them (Hay and Fenical, 1988).

Spatial escape is achieved by growing above the level of grazer action on the shore, or by settling in inaccessible crevices. Many macroalgae can survive long enough to be established only where their propagules settle in pits or cracks which grazing limpets cannot reach. Temporal escape can be achieved by growing or fruiting at seasons when grazing pressure is low.

Tolerance of grazing may involve a fast growth rate to compensate for loss of tissues so that the plants can regenerate. Fast growth, usually measured as a high production:biomass ratio, is particularly associated with microalgae because of their small size, but there is great variation within the macroalgae. Several seaweeds also have resistant phases—spores or vegetative portions—that can withstand passage through the gut of grazers and remain viable and able to regenerate.

Deterrence of grazers is the mechanism that has received most attention. The most obvious approach is shown by algae in which the whole plant—known as the 'thallus'—is modified into a particularly unwelcoming form. For instance, grazers find it difficult to penetrate deeply into the encrusting calcified thalli of some of the red algae (see p. 72). In these calcified forms, the growth layer or meristem, as well as the conceptacles (cavities containing the sex organs), are deep below the surface, so are protected from grazing. Other shapes and designs of thallus may be important in reducing herbivore effects, as shown in a comparison of the structures of opportunistic species such as *Ulva* with forms such as *Pelvetia* and *Corallina*, which tend to be longer-lived and to form more permanent members of the community. *Ulva* has high productivity, a low proportion of structural components in the tissues, and a delicate thallus susceptible to grazers. *Pelvetia* and *Corallina* show low productivity but have large proportions of structural tissues and very tough fronds that deter herbivores. These structural adaptations have been gained at the cost of a low growth rate and low reproductive output.

Structural deterrents may also be combined with chemical ones, as shown by the response of *Fucus distichus* to the grazing winkle *Littorina*. This seaweed, like other *Fucus* species, produces regenerative outgrowths if it is wounded (Fig. 4.2). These 'adventitious branches' usually grow

Fig. 4.2 The preference of *Littorina scutulata* for various parts of the alga *Fucus distichus*. Adventitious growths are shown in black (S, short; M, medium; L, long). When compared with apical meristematic tissue, snails always avoid all types of adventitious growths, probably because of their high levels of phenolic compounds. Bars show SE. (After Van Alstyne, 1989.)

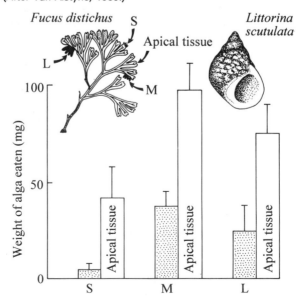

out from the midrib in bunches, and on Tatoosh Island, Washington State, on the Pacific coast of North America, their presence correlates well with the presence of grazing *Littorina*, which are known to cause damage to the *Fucus* thallus. When the plants were clipped experimentally to simulate grazing, they produced adventitious branches in 3 months: none of the unclipped plants produced them, but 62% of those clipped did so (Van Alstyne, 1989). Since preference tests showed that snails much preferred apical tissue to the adventitious branches, the adventitious branches were presumably produced to deter herbivores. They were not, however, purely a morphological deterrent. The short branches had higher concentrations of phenolic compounds than the tips of normal fronds, and overall the grazed algae contained more of these phenols than ungrazed plants.

Chemical defence may in fact make a major contribution to the deterrence of grazers by algae, and may take a variety of forms. First, most macrophytes contain a high proportion of 'structural' compounds such as lignins. These can only rarely be digested by grazers, but also mechanically hinder eating, so in a sense form a physical rather than a chemical defence. Secondly, macrophytes usually have a low nutrient content, as measured by available nitrogen and phosphorus. Most grazers require a C:N ratio of less than 17:1, whereas macrophytic algae have values above this (Fig. 4.3). In contrast, the microalgae have much lower ratios. Since herbivores may be limited by the low nitrogen content if they attempt to feed on macrophytes, they tend to feed on microalgae, which have high nutrient content and little structural material. Microalgae also have the advantage (for grazers), that they reproduce rapidly so that populations are not usually totally removed by grazing.

Fig. 4.3   Ratios of carbon to nitrogen in a variety of marine plants. Dashed outlines show the range of values. (Data from Fenchel and Jørgensen, 1977.)

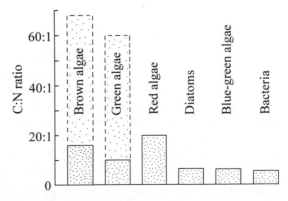

Chemical defence by macrophytes extends to much greater extremes than have so far been discussed. Brown algae produce polyphenolics, which may occupy up to 15% of the dry weight of the algae. Some of these phenols have antibiotic properties, others are antifungal, while some are 'antifeedants' and literally deter feeding. When the polyphenolics from *Fucus vesiculosus* and *Ascophyllum nodosum* were incorporated into agar discs and offered to *Littorina littorea*, the snails apparently decreased their feeding by 50% or more in comparison with those fed on normal agar, although in these experiments insufficient replication and analysis were available to make firm conclusions. Red algae produce phenols and terpenoids which may amount to 5% of the plant's weight. *Laurencia*, in particular, forms complex terpenoids which are toxic to fish and insects, and are antifeedants.

The production of a varied repertoire of defensive chemicals by algae must have a significant cost: energy that could have been used elsewhere must be diverted into the production of appropriate substances. There are no direct estimates of what this cost might be, but it is assumed from the high concentration of some of the toxic compounds produced that it is substantial.

The fate of algae and the detritus food chain

For the reasons discussed above, macroalgae die or cast off fronds, but are seldom eaten to any large extent by grazers. Instead, by far the most part of dead organic plant material is decomposed by bacteria and fungi, forming detritus. Added to this, plants produce 'exudates', such as mucus, and lose other dissolved organic material when alive, and this too is utilized by bacteria. *Any* organic carbon lost from a particular trophic level other than by direct consumption can therefore be defined as detritus, though it is a debatable point as to whether the bacteria and other microbes within the material that is breaking down should, themselves, be considered part of that detritus.

On rocky shores, the major sources of detritus will be macroalgae and microalgae. Until recently, little work has been carried out on the decay of algae. Studies on decaying angiosperms such as the tropical seagrass *Thalassia* and the saltmarsh grass *Spartina* have shown that as microbial flora builds up and the plant fragments become smaller, the carbon:nitrogen ratio falls. This is probably due to the increase in microbial protein, and it is likely that detritivores gain much of their energy from digesting this microbial fraction. However, some detritivores can also take up nitrogen from detritus without microbial involvement. The polychaete worm, *Clymenella*, for instance, can accumulate nitrogen directly from *Fucus* detritus (Mann, 1988), so the degree of involvement of bacteria in the utilization of detritus is now in doubt.

The utilization of macroalgal detritus by the benthic ecosystem has been demonstrated in a large-scale study of a kelp community on the west coast of Cape Peninsula, South Africa. The kelp beds here are dominated by *Ecklonia maxima* and *Laminaria pallida*, which produce about 500 g C m^{-2} y^{-1} in the form of particulate organic matter, and another 250 g C m^{-2} y^{-1} as dissolved organic matter. Local phytoplankton also produces about 500 g C m^{-2} y^{-1}. A simulation model has shown that suspension-feeding invertebrates could gain enough energy to explain their growth rates by feeding on a combination of detritus from kelp and on phytoplankton. During part of the year, the nearby Benguela current produces upwelling, and at this time the suspension feeders utilize detritus from kelp. When the current switches to downwelling, they receive most food supply from the phytoplankton. In this example, the detritus is essential as an energy supply for the suspension feeders at least during part of the year.

In the next chapter, the effects of grazers on intertidal algae are considered. Many of these grazers will in fact be eating a mixture of living cells and detritus. The example above which shows that animals can switch from detritus to phytoplankton as a food source supports the idea that placing animals in rigid categories of 'herbivore' or 'detritivore' is doomed to failure.

The importance of detritus as a food source on rocky shores will vary greatly with exposure. On wave-exposed coasts, most detritus will be rapidly carried away, and suspension feeders will derive most of their energy from plankton. On sheltered shores, the slower currents will allow organic material to settle near the bottom, and both suspension feeders and grazers will be able to utilize it. As we shall see later (p. 158), this difference produces striking effects on overall food webs.

Brown algae: the fucoids

Fucoid algae are found from very exposed shores to sheltered inlets and even on saltmarshes. They vary in structure from those with a classical branching ('dichotomous') form to long thin fronds. Some species form a fine turf, while in one species the reproductive parts are larger than the whole of the rest of the plant. Their morphology differs widely from one shore to another, so they provide excellent examples of how growth can be modified by environment.

Habitat and distribution

On the shores of north-west Europe there are five very common species of fucoids. Of these, *Pelvetia canaliculata* and *Fucus spiralis* are found on the high shore, *Fucus vesiculosus* and *Ascophyllum nodosum* on the mid shore and *Fucus serratus* on the low shore. In addition, on exposed shores, *Himanthalia elongata* is typical on the low shore. As one goes further north, fucoids become even more dominant, and the above species are joined by another, *Fucus distichus*, common in Norway. These species are all natives, but a recent import from

Japan, *Sargassum muticum*, is now common below low water and in tide-pools on the shores of the English Channel.

Pelvetia canaliculata and *Fucus spiralis* are remarkable in being able to tolerate extreme desiccation. *Pelvetia* may remain out of water for several days during neap tides, and in summer it can become dry and brittle, losing up to 96% of its water content (Schonbeck and Norton, 1979). *Fucus spiralis*, however, is unable to tolerate conditions in the *Pelvetia* zone. When transplanted upwards 10 cm, all the plants died within 8 weeks. This death was almost certainly due to effects of desiccation, since both *F. spiralis* and *Pelvetia* are periodically killed at the top of their range by periods of calm dry weather. Further evidence that desiccation tolerance is of the utmost importance to these high-shore weeds comes from the finding that they increase their tolerance as the summer progresses. This process of 'drought-hardening' develops in response to sub-critical desiccation.

The differences in the physiological competence of the two species are sufficient to account for differences in the upper limits of their distribution. Neither can live any *higher* on the shore than they do because they are continually 'pruned back' by effects of high temperature. Because *Pelvetia* is more tolerant than *F. spiralis* it can live a few more centimetres out of water. In contrast, the lower limits of the two algae on the shore are more likely to be set by competition and grazing, as discussed on p. 24.

Fucus vesiculosus grows most luxuriantly on sheltered shores, where it is often covered by epiphytes, but on exposed coasts it can also be common, in a form without air bladders known as forma *linearis* or var. *evesiculosus*. *Ascophyllum nodosum*, characteristic in Britain of sheltered rocky shores, can also occur as var. *mackaii*, which is an unattached form found in very sheltered sea lochs of Scotland and Ireland. On the Atlantic coast of North America, its requirements appear to be quite different from those in Europe, since it is characteristic there of exposed shores as well as sheltered bays (see p. 33). *Ascophyllum* usually harbours few epiphytes except for the red alga *Polysiphonia lanosa*.

On exposed shores, the thong-weed *Himanthalia elongata* often forms a dense belt just above the laminarian zone. Here its 'thongs' have few epiphytes. In shelter, though, they develop a 'microforest' of filamentous algae which provide a habitat for a variety of faunal epiphytes: gastropods, amphipods and the larvae of flies. The basal 'buttons' have epiphytes on their undersides, but few on their top surface, probably due to specific protective mechanisms. *Himanthalia* can also be found on some sheltered shores, but is unable to tolerate low salinities or silt, both of which affect germination and attachment of the zygotes. It is replaced in shelter by *Fucus serratus*, an alga well known for the abundance of the epiphytes that grow on its fronds.

The Japanese seaweed *Sargassum muticum* was introduced to British

Columbia, and then spread down to New Mexico. In February 1973 it was found attached to the substrate on the Isle of Wight. It spread around the Isle of Wight, and drift plants reached Normandy by 1976 and Boulogne and Plymouth by 1977. Many new sites followed on both sides of the English Channel. The invasion of *Sargassum muticum* probably occurred because it was accidentally imported with Japanese oysters into French oyster beds. This invasive alga is well fitted for rapid spread: it grows at up to 4 cm per day, survives low salinity and low and high temperatures. It has air bladders which keep fragments afloat, so that they drift. Drifting fragments continue to grow even though they can never reattach. Since they are self-fertile, they can act as centres for new zygote production.

Life histories and reproduction

Central to any understanding of seaweed ecology is a knowledge of life histories. Most macroalgae show an alternation of generations, in which the conspicuous plants found on the shore are the diploid sporophytes (Fig. 4.4a). This generation reproduces asexually, forming haploid spores which disperse, settle, and grow into haploid plants—the gametophytes. The haploid generation is generally (but not always) insignificant in appearance, but its importance is enormous since these plants reproduce sexually. Male and female gametes released from separate gametophytes combine to form a zygote which is diploid, and grows into a new sporophyte, completing the life cycle. Relatively little is known about the gametophyte generations, because of their small size. In the fucoids, however, the cycle appears in a condensed form (Fig. 4.4b). The haploid generation consists only of single cells that are formed within the reproductive organs (gametangia) of the sporophyte and these spores act directly as gametes. They fuse to form a zygote, eliminating the haploid gametophyte stage. There is, therefore, only one multicellular generation to consider in the fucoids.

The adult plants develop reproductive organs only at specific times of year. Gelatinous structures, the receptacles, develop at the tips of the fronds, and within these are cavities, the conceptacles, where spores are produced in the gametangia. The reproductive period for *Pelvetia* and *F. spiralis* is in the summer, despite the physical rigours experienced at the top of the shore. The plants are hermaphrodite, so that each individual produces both male and female spores. They also show self-fertilization, possibly as an adaptation to life high on the shore.

Fucus vesiculosus begins to form receptacles in January, and release occurs from March to August. Each plant may release something of the order of a million eggs per year, so dense settlement may occur. *Fucus serratus* reproduces in the autumn. After fruiting, it shows extensive defoliation and the fronds that bore receptacles are shed in winter, often reducing the plants to a very small proportion of their full size. Plants live for 4–5 years.

Fig. 4.4   Life cycles of (a) *Laminaria* and (b) *Fucus*. In *Laminaria* there is an alternation of generations between the large diploid sporophyte and the minute haploid gametophytes. In *Fucus*, the cycle is condensed and there is no gametophyte stage. (Partly after Scagel *et al.*, 1984.)

(a) *Laminaria*

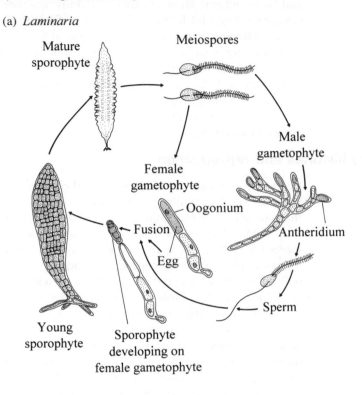

(b) *Fucus*

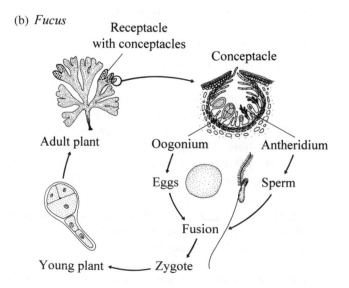

Himanthalia reproduces in July, and zygotes settle from then until late winter, growing initially into vegetative thalli or 'buttons'. By the following July, these have the form of an inverted cone attached to the substrate by its apex. From then onwards, two receptacles or 'straps' begin to grow out of the centre of the button and then branch in October. The straps reach their full length (2 m or more) by July of the third summer, when the gametes are discharged. The whole plant then breaks up. This life history is unusual in that, whereas the first year of life is devoted entirely to vegetative growth, the second produces entirely reproductive tissue, which is also remarkable in that it forms the majority of the biomass of the plant.

Growth

The zygotes of most fucoids probably settle within metres or tens of metres from the parent plants, often forming dense 'turfs'. Growth can then be exceedingly rapid. Unusually for the brown algae, the growth region or meristem of fucoids is at the apex of the fronds: a single apical cell is responsible for increase in length, while other cells at the surface divide to increase the width of the frond. This is in complete contrast to some other brown algae such as the laminarians (see p. 65).

As we have noted when discussing zonation patterns (p. 24), the growth rates of species living low on the shore tend to be greater than those living higher up. On the Isle of Cumbrae, in Scotland, *Pelvetia* living in its normal zone grew 30 mm in length in a year, while *Fucus spiralis* in its own zone grew three times as much. When young plants were transplanted down into the mid-shore, the results were striking (Schonbeck and Norton, 1980). *Fucus spiralis* grew enormously, by 200 mm. *Pelvetia* grew even more rapidly to start with, but in 6 months it stopped growing and then decayed. Later observations failed to determine unequivocally the cause of death, but a fungus whose hyphae ramified through the tissues of the alga was implicated. This fungus, *Mycosphaerella ascophylli*, is present in normal plants as well as ones that rot, so the relationship between it and its host is unclear. Whatever the cause, however, it is quite clear that *Pelvetia* cannot survive on the mid shore.

Fucus vesiculosus and *Ascophyllum nodosum* can be said to have very different 'strategies' for living on the mid shore. *Fucus vesiculosus* is a fast-growing opportunist, able to colonize cleared areas rapidly, but living only 4–5 years. Experiments in which the rock surface was cleared showed rapid settlement: a month after a square was cleared at Wembury, near Plymouth, it was covered by germlings 1.5 cm high (Knight and Parke, 1950). *Ascophyllum*, in contrast, is slow-growing, may live for 10–15 years, but often takes considerable time to recolonize bare rock. Both species, however, show amazing plasticity and the ability to adapt to an enormous range of environments.

The growth of *Fucus vesiculosus* is very variable. Tagging young plants has shown variation from 0.25 to 0.7 cm week^{-1} (13–37 cm y^{-1}) in linear growth. This is not necessarily the best way to measure growth, however, because the fronds show branching. The degree of branching varies enormously, and is correlated with exposure, so that plants are bushier at exposed sites (Fig. 4.5).

Fig. 4.5 Branching (=dichotomy) and growth in *Fucus vesiculosus*. On a moderately sheltered beach in south Devon, England, plants grow large and there are long distances between dichotomies. Vesicles are abundant. On an exposed shore in Argyll, Scotland, plants are small and bushy, without vesicles. (After Knight and Parke, 1950.)

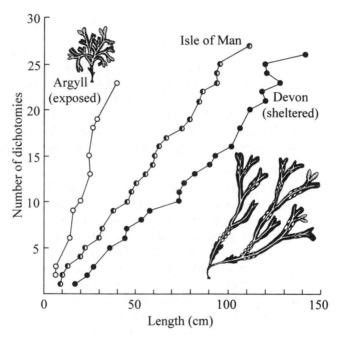

Ascophyllum has quite a different growth form from *Fucus vesiculosus*. Each plant consists of a holdfast from which arise a number of fronds. Growth of each frond is punctuated by formation of an air bladder at the tip each year, so that annual growth increments are easily measured. Absolute age is usually greater than the number of bladders, however, since the first bladder may not be formed for several years after the frond starts growing: counting bladders provides the *minimum* age. Fronds can also be broken off, and new ones may regenerate from the holdfast, so the age of the original plant may be much greater than that estimated from the bladders.

Growth and reproduction in *Ascophyllum* varies greatly in relation to exposure. On exposed coasts in Nova Scotia, Canada, fronds never become very long (up to 40 cm). Because of this, all the fronds continue to receive enough light for photosynthesis and each plant grows fast and is able to produce many receptacles. Life expectancy is short, because of wave action, and reproduction happens early (Cousens, 1986). On sheltered shores, fronds usually grow long (up to 100 cm), and these form a canopy beneath which younger plants are kept in semi-darkness. Here the young plants can grow only very slowly, and only the older ones can reproduce. Population composition on exposed and sheltered shores is therefore quite different.

The distributions of *Fucus vesiculosus* and *Ascophyllum* reflect their differing growth strategies. Because *F. vesiculosus* grows 2–3 times as fast as *Ascophyllum*, it can shade out young *Ascophyllum* plants or delay their growth. In very sheltered conditions, *Ascophyllum* can eventually grow to reproductive size, and its dense canopy will prevent settlement by *F. vesiculosus*, or by its own sporelings. Long life expectancy then allows it to dominate on these sheltered shores.

Brown algae: the laminarians or kelps

Laminarians show the typical algal alternation of generations: the diploid sporophyte is a very substantial plant, recognized as *Laminaria*. It produces haploid spores, which develop into minute haploid gametophytes (Fig. 4.4a). These produce the ova and spermatozoa, which fuse to give rise to a zygote from which grows a new diploid sporophyte. We know little about any of these stages except the adult sporophyte. Yet 'the plant' consists truly of its entire life history (Schiel and Foster, 1986). What affects spore distribution and settlement? What conditions allow, or prevent, development of the spores into a gametophyte? What affects gametophyte survival and gamete production? Under what circumstances do male gametes locate female gametes and fertilize them? All these factors are as important as, or more important than, the factors affecting directly the large plants that we can investigate at the bottom of the shore. While the following account deals mainly with mature sporophytes, we shall need to bear in mind the importance of the rest of the life cycle.

Habitat and distribution

The most extensive community inhabiting sublittoral rocky coasts of the north Atlantic is *Laminaria* forest (Dayton, 1985). *Laminaria digitata* is found on wave-exposed rocky coasts at or near the level of low water of spring tides, although it may grow deeper down as an epiphyte on *Laminaria hyperborea*. It has a smooth, flexible stipe which allows it to lie flat on the rock surface during extreme low water, so that it is protected from excessive desiccation during the brief periods for which it is emersed. The sporophyte of *L. hyperborea*, in contrast, has a stiff and roughened stipe. It is found from the bottom of the *L. digitata* zone down to variable depths. On the Isle of Man it reaches 5–19 m below chart datum, but at more wave-exposed sites in Norway it reaches –32 m. In shallow water the plants are quite closely packed, and the stiff stipes hold the blades up to the light. The rough surface of the stipes provides a good substrate for epiphytic algae and the stipe persists for a number of years, allowing them prolonged growth. It is only free of epiphytes at its junction with the blade, where growth takes place.

At wave-exposed sites the very flexible *Alaria esculenta* covers the rock at and slightly above MLWS, and in extreme wave-exposure it completely replaces *Laminaria digitata*. In shelter, *Laminaria saccharina* becomes abundant, mixed with *L. digitata*.

The fast-growing *Saccorhiza polyschides*, with a life span of only about 11 months, is an opportunist which will quickly fill any gaps in the *Laminaria* forest such as may be produced by storms. In Lough Hyne, Ireland, where it is absent from sheltered conditions, culture experiments showed that growth of gametophytes and young sporophytes was affected by the silt levels found in shelter. Transplant experiments showed that grazing on adult sporophytes by sea urchins was a restricting factor. However, the major reason for its restriction to areas of rapid water movement is that in shelter, tissue at the tips of the blades decays faster than it can be replaced.

In general, the lower limit of *Laminaria* forest is probably determined by illumination. In coastal waters green light penetrates most readily, better than blue or red, and it is the most useful for photosynthesis by brown algae. The normal depth limit of kelp forest in western Europe lies at the depth where this green light declines to about 1% of its value at the surface—about 17–20 m below chart datum.

Reproduction

Unlike the fucoids, which form reproductive receptacles at the tips of the fronds, the reproductive organs of laminarians are gathered in patches or 'sori', usually scattered over the blade. Within these are the sporangia, producing haploid spores. In most species, there is some spore production throughout the year, but peak periods are in winter for *Laminaria saccharina* and *L. hyperborea*, and autumn for *L. digitata*.

These reproductive periods are determined by illumination, as shown in experiments on *Laminaria saccharina* (Lüning, 1988). When plants were collected from the shore at Helgoland in Germany, and placed in an artificial regime of reduced daylength, they began to produce sori. Under a regime of increased daylength, no sori developed, showing that sorus production is under the control of a photoperiodic mechanism.

The number of spores produced from the sori is, to quote Chapman (1986), 'astronomical'. *Laminaria digitata* from 1 m^2 has been estimated to release 20 000 million spores in a year. The number growing to gametophytes is unknown, but in Nova Scotia, Canada, only 1 million young sporophytes were recruited per m^2 per year, and of these only two grew to visible size.

The gametophytes that arise from settled spores have rarely been seen in nature, but are well known from studies in culture. They consist of only a few cells, and male gametophytes may even be unicellular. Development in some conditions consists of vegetative growth and the production of a filamentous structure, but in others leads to rapid gamete production. There seem to be few rules that govern these alternatives, but an increase in irradiance of blue light acts as a trigger for gamete production, at least in *Laminaria saccharina*. The action of nutrients is probably important, but it is not clear which components are required to switch the gametophyte from vegetative to reproductive growth. Over a range of temperatures from 5 to 17°C, gametophytes from a variety of species became fertile in 8–20 days (Kain, 1979), so that the life span of this generation is probably very short compared with that of the sporophytes.

Growth of the sporophyte

Laminaria species produce annual growth lines in the stipe and holdfast, from which their age can be deduced approximately. Unlike the fucoids, growth takes place not at the tips of the fronds but in the transition zone between the stipe and blade, called an 'intercalary meristem'. The first pair of growth lines in the holdfast is laid down in the 7–18 month period after the plant was first formed, and thereafter each pair of lines indicates a further year of age.

The total length of laminarian plants is determined by the balance between growth (at the meristem) and shedding (at the tip of the blade). Parke (1948) punched small holes in the centre of the blade, allowing her to measure new growth (between the holes and the top of the stipe) and loss of tissue (from the holes to the tip of the blade). The blade of *Laminaria saccharina* grew in overall length from January to July, but from August onwards, the rate of shedding at the tip was greater than the amount of growth at the base, and the blades decreased in length, so that by January they were at their shortest (Fig. 4.6).

Laminaria hyperborea frequently reaches 7–12 years of age, and in the Outer Hebrides, Scotland, it may live for 15 years. *Laminaria saccharina* normally reaches 3 years, *L. digitata* 3 or possibly 5 years, *Saccorhiza polyschides* only 11 months. Juvenile sporophytes of *S. polyschides* increase greatly in numbers from June to November. They become mature from July onwards and from October onwards rot away to stipeless bulbs, which, however, continue to produce spores until the following spring, when they rot completely.

Fig. 4.6   Growth of *Laminaria saccharina* on an exposed shore in south Devon, England. Solid circles and solid line show overall length of the sporophyte. Solid histograms show length added since the previous measurement, while open histograms show length cast off since the last measurement. Early in the year, growth exceeds loss and the plant elongates, but later more is cast off than grows. At the end of the second year, cast tissue reduces the plant to a small remnant, and it dies. (After Parke, 1948.)

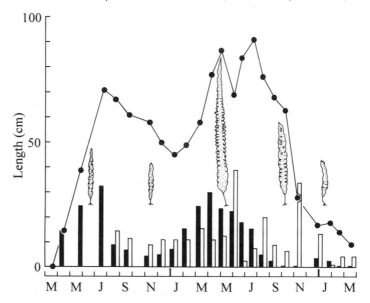

What determines the seasonal development of young sporophytes? In culture, zygotes of *S. polyschides* develop much faster than other laminarians at any temperature, and (uniquely) grow faster at 17°C than at 10°C. These properties fit in with its role as an opportunist and with its penetration furthest south, to the coast of Morocco. Development in these sporophytes in fact depends upon the amount of light that reaches them: in experimental conditions, daylength was immaterial, and even very short days with high irradiance allowed their development.

In contrast, there is evidence that the formation of a new blade in *Laminaria hyperborea* is controlled by daylength. In this species a photoperiod of 8 hours light : 16 hours darkness allowed the formation of a new blade, and so did continual darkness; but treatment with a light period longer than 12 hours prevented the formation of a new blade.

To complicate matters, studies on the growth of *Laminaria longicruris* at sites in Nova Scotia suggest that growth rate there is regulated by an interaction between temperature, light and nitrate concentration in the water. It seems that there is no one simple answer to the question of what triggers laminarian sporophytes to grow.

Variation in the form of sporophytes

In Norway, both *Laminaria digitata* and *L. hyperborea* can appear with a rounded blade instead of one with 'fingers', known as forma *cucullata*. This form of *L. hyperborea* is found in sheltered places at the lowest level occupied by laminarians, down to −30 m. Transplantation experiments in Norway showed that when the normal form was moved to a quieter site it grew a new blade that was cucullate in form. Similar transplants of the annual laminarian *Saccorhiza polyschides* at Lough Hyne, Ireland, produced a distinctly rounded (or cucullate) form in shelter, showing that the change is environmentally determined.

Wave action and competition betweeen laminarians

Competition between laminarian species probably varies in relation to wave action. In shelter, *L. saccharina* out-competes *L. hyperborea*, but it cannot endure nearly so much wave action as can *L. hyperborea*. This is well shown in the Sound of Jura, Scotland, where the lower limit of *L. hyperborea* extends further down with increasing wave-exposure, replacing *L. saccharina*.

On Amchitka Island, Alaska, the laminarian canopies extend down to −20 m (Dayton, 1975). The highest canopy consists of *Alaria fistulosa*. Below this is a canopy of *Laminaria* species, then a prostrate canopy of another laminarian, *Agarum cribrosum*, and finally below this a turf of various red algae. Experimental removal of *Laminaria* produced a dramatic increase of *Alaria* and a significant increase of the red algal turf, suggesting strong competition for space between the various algae.

The effect of sea urchins on laminarians

Sea urchins are voracious grazers on laminarians. On the Pacific coast of North America, drifting *Macrocystis* forms an important source of food for them, but urchins may also destroy living algae over wide areas (see p. 85). The effect of urchin grazing has been demonstrated in a dramatic experiment by Jones and Kain (1967) on the Isle of Man. Here, *Echinus esculentus* was removed from a strip about 10 m wide, extending from chart datum to a depth of 11 m. Before this removal there were many *Echinus* in the bottom 3 m, and no large algae at these levels. Removal was carried out at approximately monthly intervals because wandering sea urchins reinvaded the strip, and nearly 3000 urchins were removed over the 3 year period of the experiment. *Laminaria hyperborea* sporophytes then developed right down to the bottom of the strip, and in the course of the experiment grew into second- and third-year plants, demonstrating that their former absence had been due to grazing.

Green algae

The common genera of green algae are not easy to identify to species. We shall, therefore, in the main consider the greens only in terms of genera. Another confusing point about green algae is their life history. Like laminarians, they show an alternation of generations, the diploid sporophytes producing spores that grow into haploid gametophytes. In *Ulva*, *Enteromorpha* and *Cladophora*, the sporophytes and gametophytes are 'isomorphic'—that is to say, they are almost indistinguishable except for the types of reproductive bodies they produce. When examining these species on the shore it is therefore often impossible to know if one is dealing with the haploid or diploid stage. In *Codium*, the situation is different. Here there is only *one* morphological phase, the diploid sporophyte. As in the fucoids, the spores act as gametes, and there is no separate haploid phase.

The one point in common for the green algae is that in all cases, the factors that affect distribution of the species can be considered to act upon one morphological 'type' only, unlike the situation in the laminarians, where haploid and diploid plants are morphologically quite distinct.

Habitat and distribution

Enteromorpha is often found as a high-shore alga, especially in spring, but it may occur at any level on the shore, particularly associated with trickles of fresh water. *Ulva* is not usually found in such dense patches, and is often present as an epiphyte, lower on the shore. *Cladophora* is mainly present on the low shore and under overhangs.

Although *Codium* species can be found on shores of moderate exposure, especially in tide-pools, they are most common in sea loughs and sheltered bays, where the branching green plants can be abundant in the shallow sublittoral. In Lough Hyne, Ireland, the abundance of *Codium* fluctuates wildly, in some years covering an area of 500 m^2, and in others being reduced to 5 m^2 (Kitching and Thain, 1983). It is possible that at this site fluctuations relate to changes in the populations of a grazing sea urchin, *Paracentrotus lividus*: this creates bare patches that allow *Codium* sporelings to recolonize, which they are unable to do in competition with other benthic algae. Decline of *Paracentrotus* may therefore encourage decline in *Codium*, though experimental testing has not been carried out, and other factors may be important elsewhere.

Both *Enteromorpha* and *Ulva* are rapidly growing opportunists. In field experiments on the Isle of Man, clearance of limpets and macroalgae between October and March led to dense growths of *Enteromorpha* and *Ulva* by April. After the *Torrey Canyon* oil spill of March 1967, shores that were sprayed with detergents in many cases suffered almost total kills of fauna and flora. By July 1967, heavy growths of *Ulva* and *Enteromorpha* coated many of these shores, showing the rapidity with which the green algae can colonize bare surfaces. How is this rapid colonization achieved? One of the major factors here is the frequency and speed of reproduction, and we deal with this first.

Reproduction and vegetative growth

The haploid gametophytes of *Enteromorpha* produce enormous numbers of motile gametes which cluster, fuse and produce zygotes. The resultant sporophytes also produce large numbers of motile spores. Both gametes and spores can be released in such enormous numbers into tide-pools or slack water that the water mass is coloured green. As a compliment to their numbers, lumping both spores and gametes, they are termed 'swarmers'. Swarmers are often released in relation to tidal cycles: the immediate trigger may be the incoming tide as it wets the thallus, but the degree of release is usually related to the stage of the spring/neap cycle, allowing regular periodicity and synchronization of reproduction.

The gametophytes of *Ulva* release their gametes a few days before the sporophytes release their spores. In *Enteromorpha*, both types of swarmer are released at the same time. When plants of *Enteromorpha intestinalis* were collected daily from a shore at Menai Bridge and then placed under constant illumination, swarmers were released every 2 weeks as the tides rose from neaps to springs, but production fell to low levels as springs began to decline (Fig. 4.7). No experiments were conducted to assess the basis of this periodicity, but it is likely to be governed by a circa-semilunar rhythm.

Fig. 4.7 The liberation of 'swarmers' (gametes and zoospores) by *Enteromorpha intestinalis*. The plants were collected daily from a shore in the Menai Straits, Wales, and placed in dishes in the laboratory. An arbitrary ten-point scale was used to measure swarmer liberation, which peaked just before the highest tides of each neap–spring cycle. (After Christie and Evans, 1962.)

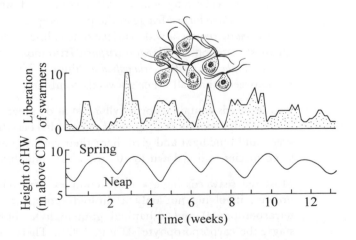

When gametes are released by *Ulva*, zygote formation and development begin quickly. Fusion of gametes may occur 3 min after pairing and zygotes settle after 5 min. Subsequent growth is very fast, as shown by the rate at which plants fix carbon. On the west coast of North America, *Ulva* fixed carbon at a rate of 10 mg C g^{-1} dry wt h^{-1}. This was an order of magnitude faster than fucoids, which in turn fixed carbon an order of magnitude faster than encrusting algae.

The success of *Codium* as a colonizer may have less to do with its method of sexual reproduction than with the ability of adult plants to disperse by floating. *Codium tomentosoides* was introduced to the coast of Long Island, New York, in 1956. It then spread rapidly north to Cape Cod (by 1961) and south to New Jersey (by 1966). The original introduction probably came from plants attached to the hulls of ships arriving from western Europe. The dispersion along the coast of North America most likely came when whole or fragmented detached plants floated in coastal currents, since the swarmer stages are not thought to be able to account for more than local dispersal. *Codium* from Europe was itself an introduction from East Asia, arriving about 1900. The species has now spread to the Pacific coast of the USA, and to New Zealand.

Red algae

The red algae show an extraordinary range of form and life style. Some are frondose and branching, some are filamentous, others are encrusting. There are annuals and perennials, free-living species and parasites. Most live relatively low on the shore, but some can live near the top of the intertidal zone—*Catenella repens* forms a moss-like turf under *Pelvetia*, while *Porphyra* ('laver') has slimy fronds that are found at all levels on exposed shores. On sheltered shores, the epiphyte *Polysiphonia lanosa* may cover much of its host plant, *Ascophyllum*. In rock pools, the pink encrusting species may achieve 100% cover, and other calcified species such as *Corallina* form a fringe at the pools' edges. Low on the shore, near the *Laminaria* region, a wide variety of frondose forms is found. Some of the most common are *Chondrus crispus* ('Irish moss'), *Laurencia pinnatifida*, *Ceramium rubrum* and *Palmaria palmata* ('dulse'). There are more species of red seaweeds than all other seaweeds combined.

Red seaweeds contain chlorophyll *a* just as do green and brown species, but they also contain red pigments called phycoerythrins, which absorb green and blue light and give them their colour. These pigments allow some species to live down to 200 m, where blue light dominates.

Most red seaweeds have a more complex life cycle than greens and browns, involving not only a diploid sporophyte (here called the tetrasporophyte) and a haploid gametophyte, but an extra diploid stage, the carposporophyte (Dring, 1982). There are, therefore, three separate sets of reproductive propagules: male and female gametes released by the gametophyte; carpospores from the carposporophyte; and tetraspores from the tetrasporophyte. Unlike the other macrophytic algae, none of these has flagella. The situation on the shore is not, in fact, as complex as it might be because the carposporophyte develops on, and remains attached to, the gametophyte, so there are only two types of 'plants' to be seen. The gametophytes and tetrasporophytes may be identical ('isomorphic') or different in structure ('heteromorphic') as in other algae. In the case of the heteromorphic species, the phases may be so different that they have been described as different species. The encrusting 'species' *Petrocelis cruenta*, for instance, is actually the tetrasporophyte phase of *Mastocarpus stellatus* (= *Gigartina stellata*), an erect frondose species.

Habitat, distribution and growth of frondose species: *Mastocarpus* (*Gigartina*) and *Chondrus*

A major factor controlling distribution of red algae is the degree to which individual species can withstand the action of waves and currents. In Maine, on the east coast of North America, the frondose gametophytes of both *Mastocarpus* (*Gigartina*) *stellatus* and *Chondrus crispus* are abundant on exposed coasts, near mean low water, but large plants of *Chondrus* suffer more dislodgement in winter than do *Mastocarpus* (Dudgeon and Johnson, 1992). The stipes of the two species are different in structure: thick and extensible but weak in *Chondrus*; thin and stiff but strong in *Mastocarpus*. Yet the force required to break the stipe is the same, for a given biomass, for both species. Why is *Chondrus* more easily detached? The answer probably lies in the structure of the fronds, not the stipe: *Chondrus* is more bushy and therefore for any given biomass a plant experiences greater drag than *Mastocarpus*. *Chondrus* is not, however, totally replaced by *Mastocarpus* because it has a higher growth rate and can regenerate rapidly from its holdfast. The two species therefore persist together in a mixed community.

The frondose gametophytes of *Mastocarpus* are annuals, recruited each spring and autumn, and they grow rapidly. The crustose tetrasporophyte plants, in contrast, are slow growing and may live for many decades, as shown at Roscoff, Brittany, where Dion and Delepine (1983) devised a technique for growing germlings of the crustose phase in the laboratory and then placing them in the field. They were able to show that growth rates of the tetrasporophyte crusts depend heavily upon height on the shore. Maximum growth rate occurs just above the laminarian zone. Growth rate is lower under the laminarians themselves, and above them at MLWN it is negligible.

The crustose and upright forms of *Mastocarpus* are therefore probably adapted to quite different ways of life. In Washington State on the west coast of North America, the crusts and fronds of *M. papillatus* have very similar distributions on the shore, and are probably subjected to similar pressures of desiccation and other physical factors. The responses of the two phases to grazing gastropods are, however, quite different. When herbivores were excluded from the crust, a thick film of diatoms and macroalgal sporelings developed and the crust disappeared. Exclusion of herbivores from the frondose phase led to establishment of *more* fronds. Grazers therefore presumably normally encourage the crusts by removing the epiphytes, but remove the more vulnerable fronds at an early stage. This may constitute an example of 'bet-hedging' by the alga: the two phases are adapted to different conditions, and should these change, one or other phase can take over. Since the sporophyte crusts can live for 40 years or more, while the gametophyte fronds are able to short-circuit the reproductive cycle and produce more gametophytes, the species can emphasize either phase for a considerable period.

Habitat, growth and reproduction of encrusting species: *Lithophyllum*

There are many species of encrusting 'corallines', which have no frondose phase, most of which are difficult to identify to species. Generally they are slow growing, but they may grow both vertically (in thickness) and laterally (at the edge of the crust). The growing region, or meristem, is usually well below the surface, so it is protected from grazers such as limpets (Steneck, 1986). Grazing is in fact usually of positive advantage in removing epiphytes, as described above for the crust phase of *Mastocarpus*. In summer, corallines may be bleached and lose their pink pigment, but in some species at least (e.g. *Phymatolithon*) this does not necessarily result in the death of the plant, and pigment may be resynthesized.

Lithophyllum incrustans is a common encrusting species on exposed rocky shores in north-west Europe. Besides having a meristem well below the surface, this species has its conceptacles (reproductive organs) buried in the calcified thallus, and connected to the exterior by canals (Fig. 4.8). These conceptacles are formed in a separate layer each year, so the layers can be used to age the plants (Edyvean and Ford, 1986), and some populations contain individuals as old as 13 years. Spore production by *Lithophyllum* is enormous (though nowhere near as high as in laminarians)—up to 18 million m^{-2} y^{-1}. In Wales, however, this enormous spore production results in the recruitment of only 55 'O class' plants—'O class' being those without a layer of conceptacles. Survival of spores is thus only of the order of 3 in 1 million, and this species has a 'low cost/low survival reproductive strategy'. Once the plants settle and have survived for a year, however, survival increases dramatically, so that the 3-year-olds form 10% of the population.

Fig. 4.8   The age distribution of *Lithophyllum incrustans* in tide-pools at Manorbier, Wales. Plants were aged by counting vertical layers of conceptacles. Most plants were young (<1 year), but some were as old as 13 years. (After Edyvean and Ford, 1986.)

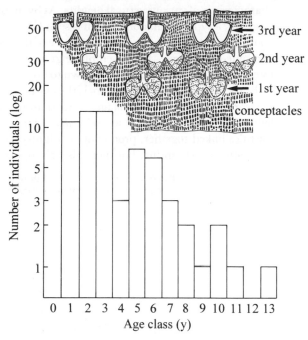

Sublittorally, corallines may grow unattached to the substrate, and are then collectively known as 'maerl'. Maerl beds consist of twig-like forms or knobbly clumps up to 10 cm across, and are not usually found below 20 m.

Microalgae

Although the most obvious plants on rocky shores are the macroalgae, and they show large standing stocks or biomass, much of the production that is important to animals may be associated with 'microalgae', principally the diatoms and blue-green algae or cyanobacteria. If production rates are of the same order as those found in the benthic microalgae of estuarine muds and sands—a few hundred grams of carbon fixed per square metre per year—they may be responsible for a considerable fraction of the total carbon fixed on rocky shores. Together with the sporelings of macroalgae and the complex of protozoa and other microbes in the 'microbial film' on rock surfaces, the microalgae provide most of the food that maintains populations of limpets, winkles, topshells and other herbivores.

Vertical zonation and seasonal changes

Just as for macroalgae, the assemblages of diatoms on rocky shores are characteristic of various vertical heights. At Swanage, on the south coast of England, a typical zonation pattern is seen (Fig. 4.9). Here the spray zone above *Fucus spiralis* is covered by a mixture of the diatom *Achnanthes brevipes* and blue-green algae. This assemblage grows in autumn and winter, and by early spring produces an olive-brown to brown layer, making the surface of the rocks very slippery. In summer, the assemblage disappears. Below the *Fucus spiralis* zone, an assemblage of the diatoms '*Fragilaria*' and *Melosira* follow the same time-scale as *Achnanthes*. These species form chains of frustules linked together, producing a bright-yellow layer of microscopic filaments in early spring, but these also disappear in summer.

Fig. 4.9   Zonation of some diatoms at Swanage, England (after Aleem, 1950). Note that the genus '*Fragilaria*' is placed in inverted commas because its present taxonomy is uncertain.

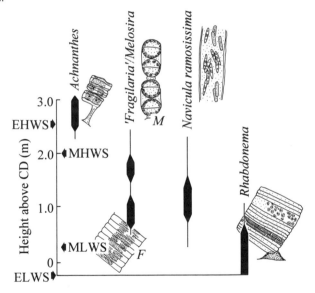

Lower on the shore, in the *Fucus serratus* zone, species of diatom such as *Navicula ramosissima* are abundant. These diatoms secrete a mucilage tube that is attached to the substrate at one end while the other end floats free in the water. Within the tube, individual diatoms can move and divide. The filaments can be mistaken for hydroids or filamentous macroalgae, but are finer and less rigid. At the bottom of the shore, in the *Laminaria* zone, diatoms are rarer. Some of the pools contain a mixture of the diatoms *Rhabdonema* and *Licmophora*, while *Grammatophora*, mixed with the green alga *Cladophora*, grows on boulders.

The composition of diatom assemblages varies not only with vertical height, but with the type of substrate. On chalk shores, many of the species that die out elsewhere in summer continue to dominate, because the chalk absorbs and retains water (Hopkins, 1964). On wood, the species common in winter give way to *Achnanthes* and blue-green algae in summer. This is probably because, while wood *absorbs* water readily, it also *loses* it readily when subjected to desiccation.

In the intertidal zone of Lough Hyne, Ireland, blue-green algae are the only abundant microalgae at the top of the shore (Little *et al.*,1990). Limpets feeding in this zone have their guts packed with rock fragments and occasional blue-green filaments. This is because some of the blue-green algae are partially endolithic—that is, they live actually within the rock between the component particles—and the limpets gouge deep into soft rock to extract them. Evidently, grazing on microalgae could have important effects, and we now turn to these.

The effects of grazers

The grazing activities of limpets and winkles depend a great deal upon temperature, so there is much more grazing pressure on microalgae in the summer. The ability of winkles to remove diatom cover almost completely at this time is well seen on the Oregon coast. Here, Castenholz (1961) used cages to enclose set numbers of winkles in areas within tide-pools, and then observed the effects on diatoms (Fig. 4.10).

Fig. 4.10   The relationship between diatom cover and volume of grazing winkles (*Littorina scutulata*) in experiments carried out in tide-pools on the Oregon coast. Lines join the outline of extreme values. One of the most abundant diatoms is a tube-dwelling form, *Navicula ramosissima*, whose tubes also contain *Nitschia* species. (After Castenholz, 1961.)

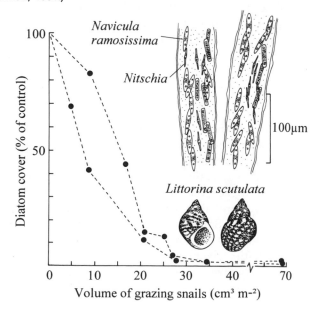

After 14 days in summer, diatom cover declined dramatically in proportion to the total volume of winkles added. When winkle volume was increased to 30 cm^3 m^{-2}, there was effectively no diatom layer. In winter, however, patchy diatom growth was observed even in areas containing many limpets and winkles.

Another way of investigating the effects of grazers on microalgae is to use artificial surfaces where numbers of grazers can be regulated. On San Juan Island, Washington, vertical clay pipes were set up in a mudflat, thus ensuring that grazers could not move across the mud from one pipe to another. On these pipes, one species of limpet, *Notoacmaea scutum*, reduced diatom density to a mean of 7 cells mm^{-2} from a mean in ungrazed areas of 304 mm^{-2}. In this case, grazed areas had consistently less algae than ungrazed areas at all seasons.

A further useful technique is to examine natural rock surfaces before and after grazing, using scanning electron microscopy. A great proportion of available biomass is often provided by *Melosira* spp. and '*Fragilaria*' *striatula*, both of which form long chains and make up a dense upper layer. Within this and beneath it are other species such as *Achnanthes* spp. and small naviculoids. Examination of the contents of limpet guts suggests that the limpets pull the chain species into their mouths like spaghetti, while the species of the lower layer are left behind because of their tight adhesion to the rock surface.

Do grazers actively select particular species of microalgae for food, or do they eat what appears in front of them? On the Isle of Man, where 90% of the diatoms consist of the genus *Achnanthes*, *Navicula* and '*Fragilaria*', Hill and Hawkins (1991) examined the gut contents of the dominant grazer, the limpet *Patella vulgata*. They found a much wider variety of diatom genera in the gut contents than they collected in small areas from the rock surface. This suggests that the rare genera are patchy, and that the limpets may selectively graze in these patches. But although some rare genera such as *Pinnularia* were apparently selected by the limpets on some occasions, there was no consistent pattern of choice.

It is important that we know more about the degree of selection shown by grazers, because even a small degree of choice of food may have marked consequences for the microbial community. On the east coast of North America, the winkle *Littorina littorea* not only reduces the standing crop of algae even at low densities, but, like limpets, it shows some selection, probably determined by the physical structure of the algal community. Some diatoms such as *Achnanthes* become *more* abundant under grazing pressure than without grazers, showing the complexity of the situation (Hunter and Russell-Hunter, 1983). We discuss some of the effects of grazers on community structure in the next chapter.

Experiments to investigate algae

The distribution of microalgae

Microalgae can be sampled easily by brushing an area of the rock with a stiff toothbrush, then rinsing the brush in clean sea water. Comparison of types on sheltered and exposed shores can be made, using a good compound microscope and examining the samples after centrifuging. If a spectrophotometer is available, quantitative comparisons of available chlorophyll at different sites can be made.

Desiccation tolerance in fucoids

By allowing various fucoid species to desiccate, and measuring weight loss, it can be shown that species lose water at similar rates. Recovery from desiccation is, however, variable, and correlates with position on the shore. Differences in desiccation rate between red, green and brown algae are marked.

Photosynthesis in macroalgae

Photosynthetic rates in macroalgae can easily be measured by placing individual plants into conical flasks filled with sea water, and measuring change in oxygen content of the water over a period of time. This can be combined with the experiment above, since one way of measuring 'recovery' from desiccation is to measure photosynthetic rate. Alternatively, the use of selective filters shows how red algae, which can utilize long wavelengths with their phycoerythrin pigments, are not much affected by a filter of wavelength 565 nm, whereas green algae show significant reduction in photosynthetic rate.

Branching and vesicle formation in *Fucus vesiculosus*

This species is found on exposed and sheltered shores, but its morphology varies greatly. Branching can be measured by selecting the longest branch, then counting the dichotomies (branches) between the tip and the holdfast, and dividing by the length. Vesicle number can be measured by counting vesicles in the length of frond between dichotomies. Both characters can be related to exposure values.

Morphology of growing regions

By taking sections with a razor, it is possible to investigate the structure of the growing regions of various macroalgae. In particular, a comparison of *Laminaria*, in which the meristem is at the top of the stipe, with *Fucus*, in which it is at the tip of the frond, is very clear. A good compound microscope and glass slides are necessary.

Measuring the age of *Ascophyllum*

Ascophyllum forms a vesicle in the thallus each year, so each frond can be aged. By counting vesicles along the length of individual plants, population structure at several sites can be compared. Note that regeneration can occur from the holdfast, so these are *minimum* ages.

Long-term clearance/colonization studies

If you can return to the shore at intervals of a few weeks, it is worth trying small clearance studies. Remove all algae from, say, a square 0.5 × 0.5 m. Then examine settlement of sporelings and/or animals such as barnacles, in this area. Use a control area in the algal canopy nearby. Keep such destructive experiments to as small an area as possible, and do not destroy slow-growing species such as *Ascophyllum*.

5 Grazers and their influences

Strictly speaking, animals that make a living by eating live algae are termed herbivores, while those that eat dead remains are detritivores. The distinction is, however, extremely difficult to make in practice because most animals have a wide range of food sources. The gastropods that rasp microfilms off rock surfaces take in microalgae, sporelings, fungi and probably algal detritus. The urchins that feed sublittorally may take both live and dead macroalgae. Here, therefore, we consider all these categories under the title 'grazer'.

Of the mobile animals on rocky shores, molluscan grazers are usually by far the most prominent, and at several localities they have been shown to exert tremendous controlling influences on the algal vegetation. Towards the lower levels of rocky shores, and in the sublittoral, sea urchins may take over as the prime controllers of algal growth. Over the whole shore, however, there are probably also significant effects caused by 'mesograzers'—animals such as amphipods and isopods, which are smaller but may occur in high densities. The effects of various grazing groups varies in different parts of the world, and broad generalizations are hard to justify. We begin, therefore, with a brief outline of the world-wide distribution of the major grazing groups.

The distribution of grazers

Various families of herbivorous gastropods have specialized in intertidal life. The most widespread are prosobranch limpets and winkles, but pulmonate limpets, topshells and neritids are also common in many countries (Fig. 5.1). Prosobranch limpets (those with gills) are found world-wide. The two largest families, Patellidae and Acmaeidae, have very different distribution patterns. In north-west Europe, both families are represented, but the patellids (e.g. *Patella vulgata*) are by far the most abundant intertidally. The same is true for southern Africa, where there are 11 species of patellid. In Australia and New Zealand there are many patellids, and the most common Australian limpet, *Cellana tramoserica*, is a patellid. The patellids are, however, joined there by many acmaeids. In America, on both Atlantic and Pacific coasts, acmaeids are dominant, and in California there are 17 species of this family.

Fig. 5.1   The occurrence of some molluscan grazers.

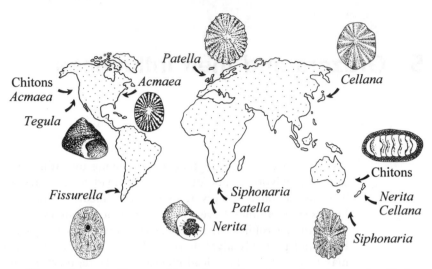

The pulmonate limpets (those breathing through a lung—the Siphonariidae) are more abundant in tropical than temperate waters. They are well represented in Australia, New Zealand, southern Africa and South America, but are not common in North America and are absent from north-west Europe.

Winkles (Littorinidae) are a ubiquitous feature of rocky shores, especially at high tidal levels. In north west Europe there are several common species, and three of these—*Littorina saxatilis*, *L. obtusata* and *L. littorea*—are also found in north-east America. *Littorina littorea* eats soft macrophytes while *L. obtusata* eats fucoids, but elsewhere in the world the littorinids seem to be abundant on 'bare' rock, eating microalgae. Topshells (Trochidae) are not usually very influential on algal growth, although they are a feature of most continents, often low on the shore. They are not common on the east coast of North America, but on the Pacific coast, *Tegula funebralis* is one of the most common grazers.

Neritids (Neritidae), like pulmonate limpets, are primarily a tropical group, but extend to southern Africa and to Australasia. They are not found in North America other than from Florida to Texas, and they have not reached European coasts despite the fact that there is a closely related European freshwater species.

The chitons (Class Polyplacophora) are the only other molluscan group apart from the gastropods to be considered as important grazers. They are distributed in all seas, but the majority live low on the shore or in the sublittoral. In Europe and north-east America, it is unlikely that they have a significant effect on the flora. In Australasia, Africa and on the Pacific coast of America, however, chitons are abundant on open rocky coasts, and probably compete with gastropods for microfloral food.

Little is known of the effects of mesograzers on most shores. Isopods of the genus *Ligia* are common world-wide, and a similar genus, *Ligyda*, is found on Pacific American coasts. Both act mainly as detritivores. Amphipods are universal and very diverse, but they have attracted little attention despite their probable importance as macrophytic grazers.

The influence of sea urchins probably varies greatly with species and locality, and urchins are found low on the shore and sublittorally throughout the world. On both Atlantic and Pacific coasts of North America, sublittoral urchins have been shown to have dramatic effects on kelp beds. In Ireland, *Paracentrotus* has a great influence on algal growth, but there is little evidence that *Echinus* has such effects. In Australia and New Zealand, urchins are common but their influences have not been investigated.

There are probably many other animal groups that act as grazers in some localities. In the tropics, fish are very influential, and at some sites crabs may graze intertidal algae. We now take some examples to show the effects that grazers may have on the shore.

The effects of grazers

One of the points that recurs in the discussion of invertebrate herbivores is the effect that they have on plant distribution. Do herbivores determine community structure? Or do they fit in to a complex network of interactions on the shore? To simplify discussion, we shall start with animals primarily thought of as intertidal and then move to the subtidal urchins. There is, in fact, no clear dividing line, since some limpets such as *Patella aspera* occur subtidally, and some urchins, such as *Paracentrotus lividus*, can be important grazers intertidally, but in general terms the distinction is a convenient one.

Intertidal grazers

An experiment on the Isle of Man, in which limpets were removed from a 10 m wide strip of semi-exposed shore, demonstrates clearly the effects of grazing. Initially, the community on this shore was dominated by limpets and barnacles, and there were few macroalgae. In the winter, all limpets more than a few millimetres long, and all large algae, were removed (Jones, 1948), so that by March the strip was bare. Within a few weeks the strip became thinly covered with a film of filamentous green algae and diatoms. Other vegetation followed, including *Enteromorpha*, *Ulva* and *Porphyra* and by July the brown alga *Fucus vesiculosus* was abundant and *Fucus serratus* occasional. Many other algae had arrived. The strip stood out dramatically as a dark band across the limestone, which was otherwise nearly bare apart from barnacles (*Semibalanus balanoides*).

After 3 years' growth, *Fucus serratus* was plentiful on the lower part of the strip and *F. vesiculosus* on the upper part, but both occurred at all levels. *Fucus spiralis* formed no distinct zone but was present in small quantities at all levels. Thus the grazing of limpets, not the mechanical action of waves, had previously limited the population of fucoids: while limpets seldom eat adult macroalgae, they are efficient and rapid grazers of the sporelings, and can prevent recolonization once a shore is cleared. On more sheltered shores, however, limpets do not affect fucoid growth so much, and the algae form dense zones. The reasons for this are unclear.

The grazing of algae by limpets was further illustrated following the *Torrey Canyon* oil spill in England (Fig. 5.2). On 18 March 1967 14 000 tonnes of Kuwait crude oil was spilled into the sea west of Cornwall, and was cast up on to the beaches in the following 10 days. These beaches were treated with 10 000 tonnes of toxic dispersants. Most animals on the shore were killed, including many limpets. There followed a settlement of *Enteromorpha* and *Ulva*, leading to a 'greening'. During the late summer and autumn of 1967, *Fucus vesiculosus* var. *linearis* (= *F. evesiculosus*) began to appear, and in places *Laminaria digitata* and *Himanthalia* grew 1.5–2 m further upshore than normal. Animal life was still much reduced in 1968, but a few barnacles settled in areas free from *Fucus*. *Patella* settled and removed much *Fucus* in 1972–73, and there was a return to the *Patella*–barnacle community by 1973–74.

Fig. 5.2   Recolonization of flat rock surfaces at Porthleven, Cornwall, after pollution by oil from the *Torrey Canyon* and use of dispersants in spring 1967. Stipple shows initial 'greening' by *Enteromorpha* (% cover not known). Solid circles show the rapid colonization by *Fucus* spp. Open circles show the slower growth of *Patella* spp., early individuals being smooth and fast-growing while later ones were rugged and slow-growing. This shore took at least 7 years to return to anything like its former state. (After Southward and Southward, 1978.)

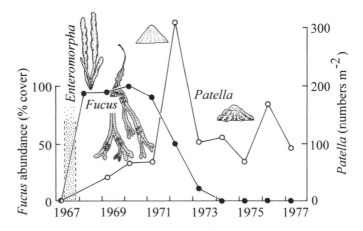

These two examples illustrate the controlling action of limpets over intertidal algae, as well as the long time periods necessary for intertidal communities to recover from gross disturbances. Interactions between grazers and algae are, however, often quite complex, and dependent upon the presence and actions of other members of the community, as shown in studies of the winkle *Littorina littorea* in tide-pools on the coast of New England. This species settles from the plankton as baby snails in high-level tide-pools, and does not appear to migrate between pools, so that each can be treated as an isolated habitat. *Littorina* prefers to eat soft green algae such as *Enteromorpha* sp., and avoids the tough red alga *Chondrus crispus*; in experiments when it was added to pools covered by *Enteromorpha* it virtually removed this species (Fig. 5.3). Conversely, when *Littorina* was removed from pools dominated by *Chondrus*, the pools became dominated by *Enteromorpha*.

Fig. 5.3   The effect of grazing by the winkle *Littorina littorea* on the algal flora of high-shore tide-pools at Nahant, Massachusetts. Experimental treatments began in April 1974. In the control pool, *Chondrus crispus* remained dominant. When *Littorina* was added to a pool dominated by *Enteromorpha*, this declined but was not replaced by *Chondrus*. When *Littorina* was removed from a pool, *Chondrus* declined and *Enteromorpha* took over. (After Lubchenco, 1978.)

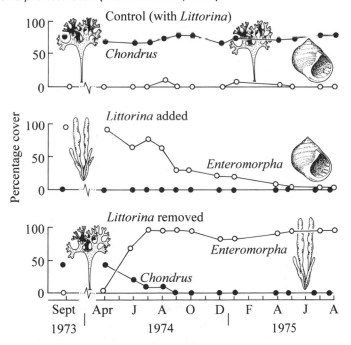

The situation in the pools is, however, more complex than a simple grazer–alga interaction. The shore crab, *Carcinus maenas*, is a predator of young *Littorina*, and gulls in turn prey on *Carcinus*. In the pools dominated by *Chondrus* there are no crabs, but in those with a thick covering of *Enteromorpha* they are abundant. There they can hide from gulls, and they eat small *Littorina* settling from the plankton. The system is self-regulating, because the *Enteromorpha* is now protected from the grazing winkles.

Lower down the shore in New England, the general rock surface around tide-pools is heavily covered with *Fucus vesiculosus*, while pools are free from this but contain a variety of other algal species. It is possible to get *Fucus* to colonize the pools, but this can only be done by excluding grazers and removing other macroalgae at the same time (Lubchenco, 1982). *Fucus* will not colonize if the only treatment is to exclude grazers, or if the macroalgae are removed without removing grazers. Thus we can conclude that the normal absence of *F. vesiculosus* from pools is brought about by a combination of grazing and competition, and not by the physical effects of continuous immersion.

On much of the open shore in New England, this interaction between grazing and competition is clearly seen. *Littorina* grazes *Enteromorpha* effectively. However, it is not effective at clearing rocks with pits and crevices, and these act as refuges for algal sporelings. *Fucus vesiculosus* plants that settle in these refuges quickly grow to a size at which they are no longer vulnerable to grazing. Since *Littorina* removes the *Enteromorpha*, it effectively aids the growth of *F. vesiculosus* by removing its competitor.

Despite its importance as a grazer, *Littorina littorea* is not in fact a native of North America, and was introduced from Europe in 1861. How much has it modified the native flora and fauna since it arrived? This question has been examined by excluding the winkle from areas of a beach on Rhode Island (Bertness, 1984). The site is sheltered and is completely covered by cobble stones. Large pens were set up, from which all *Littorina* were removed, and the effects of this treatment on the rest of the flora and fauna were observed. Removal of *Littorina* was followed by a rapid growth of the green algae *Enteromorpha intestinalis* and *Ulva lactuca* and by accumulation of sediment. The sediment deterred settlement of the barnacle *Semibalanus balanoides*, but led to a considerable population of the burrowing polychaete worm *Polydora ligni*. Much of the sediment was bound to the rock surface by *Enteromorpha*. This experiment suggests that before the invasion of *Littorina littorea* much more of the shore was covered with sediment, and with associated fauna. The present 'clean' appearance of the cobbles is due indirectly to the presence of *Littorina*.

On any one shore, the situation is thus likely to be quite complicated. Individual grazing species may influence the distribution of particular algae, but subsequent growth of algae will be determined by a mixture of competitive abilities and tolerance of physical factors. The grazing pressure exerted will, in any case, be determined or modified by predation pressure as well as by the success with which the grazer can deal with its own competitors, and the physical rigours of the environment. In many cases the grazers may be maintained below the carrying capacity of the shore by the action of predators. In summary, the grazer–herbivore link can be a very important one, but its importance can only be viewed in the context of the entire food

web. The constituent organisms of this web will show different reproductive success and survival from year to year and place to place, and
this makes it imperative that conclusions are drawn for the shore on
which experiments were carried out, and are not generalized too much.

In different geographical areas, the effects of grazing do vary enormously. In the north-east Atlantic, patellid limpets appear to be effective
in removing macroalgae from exposed shores, but not from sheltered
ones. In other parts of the world, grazing by littorinids, limpets and
chitons shows less obvious relations to wave exposure. On the Pacific
coast of North America, there is great variation in algal cover with
latitude: algae become progressively less common in the mid shore as one
moves further south. Whether this is due to grazing pressure, desiccation
or both, is not certain.

A further complication is presented by the patchy distribution of algae.
On semi-exposed shores in north-west Europe, *Fucus vesiculosus* grows in a
mosaic pattern, often as var. *linearis*. How do these patches arise and how
are they maintained? We postpone discussion of this topic to Chapter 8.

Subtidal grazers

Sea urchins are voracious herbivores. But are they in fact responsible for
controlling growth of algae in the subtidal? Observations in Nova Scotia
certainly showed that urchins *can* destroy kelp forests (Breen and Mann,
1976). The forests consist of a dense coverage of kelp (*Laminaria longicruris*
and *L. digitata*) from near low water to a depth of about 20 m. For the
most part the urchins (*Strongylocentrotus droebachiensis*) are unable to climb
these kelps and to hold on in the face of wave motion, although
occasionally there are so many urchins climbing a *Laminaria* that they
hold it down and devour it. Over the period 1968–74, 'wave fronts' of
large sea urchins moved forward, devouring the kelp at up to 1.7 m per
month, producing 'barren grounds' without any kelp at all. On these
barren grounds the sea urchins grew considerably less per year, both in
diameter and in gonad weight, than did well-fed urchins eating the kelp.

Sea urchins have also been shown to destroy kelp forests on the Pacific
coasts of America. *Strongylocentrotus purpuratus* and *S. franciscanus* removed
the *Macrocystis* forest on an offshore rocky reef in one year. Experiments
in Newfoundland in which sea urchins have been removed totally have
also shown that algal forests can regenerate rapidly, though type of
community and speed of growth varied from place to place.

If sea urchins *can* destroy kelp forests in this manner, why are there any
kelp forests left? One possible answer is that predators and other
controlling influences such as disease keep urchins down in numbers
so that kelps can always find a refuge somewhere. We discuss this further
on p. 101. Reproductive success of urchins may also be controlled by

conditions experienced by the planktonic larvae, and we discuss the topic of 'supply-side' ecology on p. 177.

A further possibility is that barren grounds and kelp forests represent part of a large-scale mosaic related to normal cyclic events. Briefly, if ecosystems cycle through a series of successions, and if these successions are out of phase in different places, we would expect to find a rather patchy environment with each patch in a different stage of the succession. We discuss this theory in detail on p. 164.

We now go on to discuss in more detail the biology of various grazers.

Limpets

In north-west Europe the limpet *Patella vulgata* is widespread on rocky shores, from sheltered to wave-exposed, at all intertidal levels. *Patella aspera* is found on the low shore and in the shallow sublittoral region. *Patella depressa* is restricted to wave-exposed low-shore situations in the south-west. There are a few other limpet species in north-west Europe, including the blue-rayed limpet *Helcion pellucidum* (found attached to various algae) and low-shore acmaeid species. Limpets are widespread around the world—the limpet form is a very efficient and adaptable living structure. Limpet radulae can be used for grazing on microalgae, on thin 'ephemerals' (short-lived algae such as *Ulva*) and on kelps. Limpet shells are conical and strong, resisting wave attack and predators. The limpet foot can adhere powerfully to the rock surface, using two separate mechanisms for attachment: suction, used mainly at high tide, and 'glue-like' adhesion, used mainly at low tide. Remarkable tenacity—up to 0.23 MN m^{-2}—can be achieved by glue-like adhesion.

Sex change, reproduction and growth

Patella vulgata regularly undergoes the remarkable phenomenon of sex change. Small individuals (about 20 mm in shell length and about 1 year old) are predominantly male, the remainder being neuter or female. Larger and older individuals are progressively more female. *Patella vulgata* is therefore referred to as a 'protandrous hermaphrodite', protandrous meaning 'male first'.

Limpets spawn by liberating their eggs and sperm directly into the sea. At most sites, spawning reaches its maximum between early September and late October, and dies back to a minimum by midsummer of the following year. The onset of spawning may possibly be correlated with high winds and rough weather. After a period in the plankton, the larvae settle as 'spat' low on the shore or in damp crevices, and as they grow, slowly make their way up to varying levels on the shore.

Their growth rate varies considerably, and appears to depend both on level on the shore and on the type and density of the associated fauna, such as barnacles and mussels. Growth is slow at high level with barnacles, somewhat faster at mid shore with barnacles, considerably faster at mid shore level with mussels and barnacles (where conditions stay damp during emersion) and very fast on low-shore bare rock. The greatest life span on low-level bare rock is estimated at 4–5 years, and on high-level barnacle-covered rock as 12–15 years or longer. Whether this relates to varying predation levels has yet to be established.

Limpets appear to lay down extra shell only during submersion. The shell shows micro-growth check lines in the shell that correspond in frequency to tidal emersions (Ekaratne and Crisp, 1982). From the number of covering tides calculated for a position on the shore, limpet age can therefore in principle be calculated exactly. However, the technique involves sectioning the shell and is quite complex, requiring the use of a scanning electron microscope.

Feeding

Patella has a shovel-shaped radula referred to as 'docoglossan'. This has, in each row of teeth, four laterals with hardened tips, two marginals with hardened tips, and six marginals without hardening (Fig. 5.4). The tips are hardened with goethite (iron oxide), and can dig into rocks as hard as limestone. Feeding occurs in regular patterns as the head is swept slowly from side to side, each rasp with the radula making a clear patch on the rock. Limpets take up microalgae such as diatoms, macroalgal sporelings, and a variety of other items (Hill and Hawkins, 1991). Since mucus trails left by crawling limpets provide a substrate for growth of bacteria and microalgae, it has been suggested that grazing of trails could benefit the limpets. The mucus of many other grazers acts similarly, and these species may possibly be regarded as 'gardening' their environment: they promote the growth of their food supply in a similar fashion to human gardeners.

Fig. 5.4 Part of the radula of the limpet *Patella vulgata*. Tips of the teeth (shown black) are hardened with iron oxide. On the right the radula is seen in plan view; on the left in side view, as it would appear when rasping on the substrate.

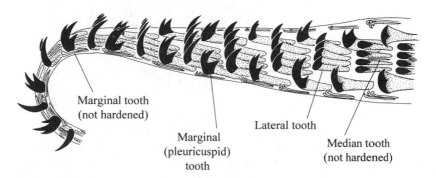

Marginal tooth
(not hardened)

Marginal
(pleuricuspid)
tooth

Lateral tooth

Median tooth
(not hardened)

Gardening of a different type certainly exists in some limpet species. In South Africa, *Patella longicosta* has a specific association with the alga *Ralfsia verrucosa*: a defined area around the adult limpet is grazed, colonized by *Ralfsia* and defended against intruders. The limpet grazes away *Ralfsia* at the same rate that the alga grows, and eliminates other algal species. When limpets are removed from their gardens, *Ralfsia* flourishes initially, but is then either removed by other species of limpet or overgrown by other algae. Note that although the biomass of *Ralfsia* is lower in the gardens than outside, the *productivity* is higher, so gardening keeps the limpet well supplied with food.

Homing

Many limpet species demonstrate the ability to home: each individual has a home 'scar', a place on the rock where the shell fits the rock profile exactly, and to which it returns after making feeding excursions. In some species, homing ability is accompanied by territoriality: on the Pacific coast of North America, *Lottia gigantea* has a grazing territory that it defends against other grazers, like some South African species.

The three *Patella* species in northern Europe usually home, although the degree of homing varies with age and with habitat. While grazing, *Patella vulgata* moves slowly, but while travelling out or back between home and the grazing area it moves faster, reaching 1.5 cm min^{-1} (Cook *et al.*, 1969). Limpets do not necessarily return home by the same track (Fig. 5.5), so how do they navigate? So far, experiments have mostly determined which methods the limpets do *not* use, and there is as yet no conclusive answer to this fascinating problem.

Fig. 5.5 Movements made by one individual of *Patella vulgata* on three different nights. Circles with stipple show the regions where grazing was prolonged. While the return tracks often follow the outward ones, this is not always so. (After Chelazzi *et al.*, 1994.)

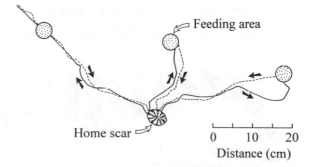

They cannot use a memory of outward movements or of topography, because individuals of *Patella vulgata* knocked off their rock and moved by some centimetres to one side of their original positions continue to home successfully. Similar experiments on the pulmonate limpet

Siphonaria normalis in Hawaii showed that limpets removed from the rock and put back in different positions on areas where they had been before, all returned home directly. In contrast, limpets placed in areas they had *not* been seen to visit did not return home (Cook, 1969). These experiments involved small numbers, but their only plausible explanation seems to be that the limpets use chemical trails, and removal of chemicals by scrubbing the rock with a wire brush does stop homing by some individuals. The use of trails may seem unlikely because, while return paths may follow outward ones, they often do not. However, it has now been shown that limpet mucus can persist on the rock surface for weeks, so that limpets may be able to use old trails as guidelines. Trail-following is well-known in a variety of other molluscs, for example chitons. How does each individual know that it is using its own trail and not that of another individual? How does it know which way to go along a trail? These are questions for the future.

The timing of foraging activity

Different limpet species are known to feed at different times in relation to tidal and day–night cycles. Some feed under water, some out of water, and others feed as the tide is rising or falling over them. *Patella vulgata* appears to be more variable in its behaviour than most. For instance, in some populations of *Patella vulgata* on the Isle of Man movements are restricted to daylight during submersion. On the other hand in the Channel Islands, movement of *Patella vulgata* is largely in the dark and while the rock is emersed.

Differences in behaviour pattern are not, however, determined by geographical locality, since in some places such as Lough Hyne, Ireland, both patterns of activity exist (Little *et al.*, 1991). In general, limpets on steep rock faces at Lough Hyne show major foraging activity at night-time low tides, although low-shore individuals also show some activity at the time of low tide during the day. There is almost no activity at high tide (Fig. 5.6). Activity on near-horizontal faces shows a complete contrast: individuals forage under water during daylight, and are relatively inactive at night. The difference does not hold at all sites, however, and the reasons for site-specific differences are at present obscure.

This behaviour is governed to some extent by an internal clock, as shown by experiments at Menai Bridge, Wales. Rocks with *Patella vulgata* on their home scars were fixed under a floating pontoon just below the sea surface, so that the animals were kept submerged although in very shallow water. The limpets all began to graze 1–2 hours after the flood tide had set in, although they were given no apparent cue in terms of rise in water level, and they returned home as the time of low tide approached.

Fig. 5.6   Activity records of two groups of *Patella vulgata* over five consecutive 24 h periods, at Lough Hyne, Ireland. Solid circles show limpets on near-horizontal surfaces. Open circles show limpets on near-vertical surfaces. Horizontal lines show when all limpets were submerged, while dots show that only some individuals were submerged. The bar shows night (black), day (white) and dawn and dusk (stipple). The near-horizontal group was active twice a day, under water, while the near-vertical group was active out of water for one period, at night. (From unpublished observations by C. Little.)

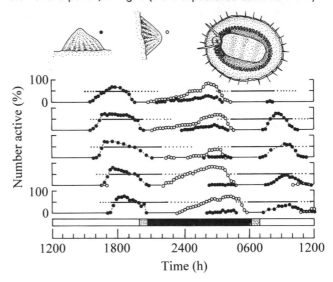

Why does the clock trigger activity at different times in different populations? Minimal activity during daytime low tides suggests that limpets are avoiding the influence of desiccation, while inactivity at night-time high tides suggests an influence from crab predation, since crabs are mainly active at night. Modelling of the time-partitioning of foraging in *Patella vulgata* supports this idea: the limpets' 'strategy' is to mimimize the time spent away from the home scar, showing that the most important selection pressures are those that act when they are away from home—probably predation and desiccation. We discuss these two factors below.

Desiccation of limpets

How important is desiccation for limpets? When rates of desiccation are measured in the laboratory, *Patella vulgata* from high on the shore loses water only very slowly, but 50% of individuals die when they have lost 60–65% of their body water. *Patella vulgata* from low on the shore loses water faster, and 50% of them die at the lower total loss of 50–55% body water. In this species, individuals therefore seem to be able to adapt to external conditions in some way. However, the situation is complicated by the fact that high-shore limpets grow much larger than low-shore limpets. The large limpets have a lower ratio of surface area to body volume and this will itself reduce loss rates. Species that live at the bottom of the shore, such as *Patella aspera*, have even greater rates of evaporation, and die when they have lost only 30–35% of their body water.

There is a further complication when considering water balance as a whole: it takes time for water to be replaced by diffusion back into the tissues from the outside sea. For a limpet high on the shore the tide may not stay high long enough to achieve this, and the limpet will start the next day of desiccation already seriously depleted. Continuous emersion for several days can therefore pose a serious threat to limpets, and some high-shore populations are killed in summer.

Anti-predator behaviour

Birds, crabs, starfish and whelks may attack limpets. It might be thought that limpets are defenceless in this situation, and when birds attack the only possible defence is to clamp down on the rock. Responses to whelks and starfish are, however, truly astonishing. South African patellid limpets respond actively to such attacks, elevating the shell and then suddenly smashing the edge of the shell down upon the predator—behaviour called 'stomping'. Stomping may seriously damage, and deter, the predator. The preliminary elevation of the shell—known as 'mushrooming'—may also dislodge predators such as starfish. Both reactions have now been observed in *Patella vulgata* (Fig. 5.7). Together with the aggressive territoriality of American and South African species, these reactions make it clear that limpets have far more capacity to react to circumstances than passive retreat into their shells.

Fig. 5.7   The response of *Patella vulgata* to a starfish, *Asterias rubens*. In normal grazing, the limpet body is hardly visible. In response to *Asterias*, the limpet raises its shell above the substrate ('mushrooming'), then rocks it from side to side. The shell is then sometimes brought quickly down to the substrate ('stomping'). (From photographs by C. Little.)

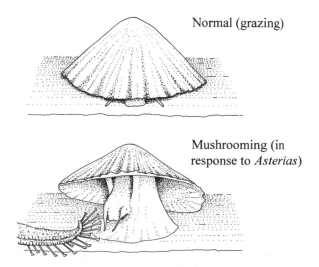

Normal (grazing)

Mushrooming (in response to *Asterias*)

Winkles

Winkles (family Littorinidae) are a feature of almost every rocky shore in the world. In northern Europe we can consider them to fall into four groups. In the first group is *Melarhaphe* (= *Littorina*) *neritoides*, a small, shiny black snail found at the top of exposed shores. In the second is the large edible winkle, *Littorina littorea*, abundant on a wide variety of less exposed shores. On weed-covered shores is the third group, the flat winkles. There are two of these, *Littorina obtusata* and *L. mariae*. Lastly, almost every shore will have some representatives of the rough winkles. The systematic status of these has been in flux for many years, but there are probably three species: *Littorina saxatilis*, *L. arcana* and *L. nigrolineata*. Reference will be found to '*L. neglecta*' in much past literature, but it is now thought that this is a dwarf form that can be adopted by any of the rough winkles (Reid, 1993). Differentiating between species is often a specialist occupation, and even the winkles themselves find it difficult: a high proportion of interspecific mating has been observed. The family Littorinidae is evidently extremely adaptable, and has produced from a basic stock a wide variety of shell forms and reproductive systems: some have pelagic larvae, some have benthic egg masses, and some brood their young. Even snails that look very similar on the outside may therefore have different ways of life, and this makes the study of littorinid ecology particularly challenging.

Shell shape and colour

The controversy about littorinid taxonomy is well exemplified in the discussion about whether *Littorina saxatilis* should be split into two species, *L. saxatilis* and *L. rudis*. On rock cliffs that are exposed to wave action, the shells of *L. saxatilis* have a relatively large aperture and are thinner than shells found on more sheltered shores. The wide, thin shells were classified as *L. saxatilis* and the narrow thick shells as *L. rudis*. However, there are intermediate shells, and it is probable that the two so-called species represent the ends of a continuous gradient of shell characters, or 'cline'. The relationship of shell aperture and thickness to wave exposure allows better adhesion in response to severe wave action (i.e. with a larger foot) and greater resistance to predators such as crabs (i.e. with a thicker shell) at the sheltered sites. Similar relationships have been shown in other littorinids (Raffaelli, 1982).

The rough winkles (*L. saxatilis* group) show great variation in shell colour. One reason for this might be to make shells less conspicuous to predators. On the coast of Wales there is a strong positive correlation between red rock and red shells of *L. nigrolineata*, especially in shelter. With increasing wave-exposure the rock becomes covered with white barnacles, and it may be that white snail shells are less likely to be noticed here by crabs or birds. The shells of *Littorina saxatilis* are also red more frequently on red rock, but the frequency of red morphs *decreases* towards increasing shelter. A possible explanation is that in shelter the increased fucoid cover hides the shells from predators, and red camouflage is unnecessary, but this idea needs testing.

The flat winkle *Littorina mariae*, common on *Fucus serratus*, also has a number of shell colour morphs. One of its major predators is the fish *Lipophrys (= Blennius) pholis*, the shanny, which eats juvenile snails. In different situations, different colour morphs provide protection against being eaten (Reimchen, 1979). The yellow morph stands out against the *Fucus* stem when seen by reflected light, and here the shanny takes it first. When it is on the transparent lamina of the *Fucus*, it is camouflaged because the lamina appears yellow by transmitted light, and the shanny ignores it (Fig. 5.8). For the 'dark reticulata' morph the converse applies. The importance of this camouflage in the wild has yet to be assessed.

Fig. 5.8   Predation by the shanny, *Lipophrys pholis*, on two colour morphs of the winkle *Littorina mariae* found on *Fucus serratus*. The morphs are 'citrina' (yellow—shown as C) and 'dark reticulata' (dark reticulations on a brown colour—shown as DR). When the light shines through the *Fucus* frond, the citrina morph is well camouflaged (to the human eye), while the dark reticulata stands out. The shanny eats more of the dark morph in this case. In reflected light, or near the base of the plant, the dark morph is cryptic while the citrina stands out. In this case the shanny preys mostly on the citrina morph. (From information in Reimchen, 1979.)

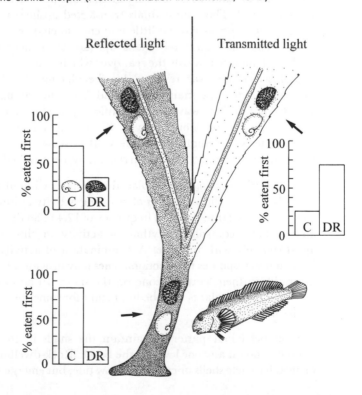

Reproduction

There is no obvious correlation between reproductive mode and shore habitat in the littorinids. *Melarhaphe neritoides*, although it lives high on the shore, has planktonic egg capsules. Both *Littorina obtusata* and *L. mariae*, living at different levels, are oviparous, laying benthic egg capsules. *Littorina saxatilis*, often found at the top of the shore, is ovoviviparous (i.e. broods its young); but *L. arcana*, also found at high levels, lays benthic egg masses. The modes of reproduction are probably better related to microhabitats, and it is even possible that some of the present 'species' are really reproductive morphs. Thus, '*L. neglecta*' is a morph of several species which is specialized for maturation at a small size, allowing life in dead barnacle shells. *Littorina arcana* could be a morph of *L. saxatilis*, specialized for exploiting the crevices of exposed shores where eggs do not dry out (Ward, 1990).

Zonation and behaviour

How do *Littorina* spp. find their way to their appropriate zones on the shore, while still moving around to feed? Although we can suggest several possible mechanisms, the answer to this question is unclear. At Woods Hole, Massachusetts, Gendron (1977) transferred *Littorina littorea* upshore and, as a 'control', lifted snails up and replaced them at the same level. Those individuals transferred upshore migrated downwards while controls showed little tendency to change their level on the shore. When he transferred snails upshore on a moonless night they could still migrate towards the sea, over which there was a brighter sky, and this direction was reversed by the placing of a lantern on the shoreward side. But snails which had been moved upshore also responded positively to waves set up artificially in a long narrow tank: they moved into the oncoming waves. Since waves move shorewards, a response to waves might guide the winkles seawards. Thus it appears that responses to light and to currents may be involved.

When *Littorina nigrolineata* was placed in aquaria without environmental cues, the winkles showed peaks of crawling activity at intervals of about 13 h—very similar to the tidal frequency of 12.4 h. So there appears to be an internally controlled rhythm of activity in this species, cycling approximately with the tide. A combination of activity rhythms and behavioural responses to directional cues may provide a basis for keeping littorinids in their 'correct' zone on the shore. In addition to this, the detection of food plants may be important for some species, as discussed in the next section.

Further behaviour patterns maintain the snails in their appropriate microhabitats at any one level on the beach. Rock-dwelling species retire to dead barnacle shells or crevices at low tide, but emerge to graze on the rock at high tide. Epiphytic species may retire to the depths of weed masses at low tide. The abundance of these 'refuge' microhabitats therefore determines, to some extent, the population density of winkles, as shown experimentally by the drilling of holes to form artificial crevices in boulders. These greatly increased the populations of *Littorina saxatilis* and *Melarhaphe neritoides*. Presumably they provided protection from both desiccation and predators. Certainly in some species, such as the Australian *Littorina acutispira*, individuals that fail to find protective crevices show higher mortality.

Relationships with food plants

Littorinids show definite preferences for particular algal foods. *Littorina littorea*, for example, prefers the green algae *Ulva* and *Enteromorpha* to the fucoids or to perennial red algae. *Littorina obtusata* and *L. mariae*, in contrast, strongly prefer fucoids (Norton *et al.*, 1990). Littorinids can also respond to chemical exudates from different species of algae. *Littorina littorea* was actually repelled by exudates of *Ascophyllum*, while *L. obtusata*, which lives on this species, was strongly attracted to it. Whether perception of such chemical cues on the shore could aid the snails in maintaining their appropriate position is more difficult to say. However, algae such as *Ascophyllum* are often present in enormous quantities, and their chemical signals may, in consequence, be very strong.

The species that live as epiphytes on algae often show intimate relationships with their host. *Littorina obtusata*, for instance, lives on *Ascophyllum*, grazes into the alga to obtain a food source, and lays egg masses on the algal fronds. *Littorina mariae* lives on *Fucus serratus*, but in contrast grazes mainly on epiphytes, not on the macroalgal tissue itself, and in so doing it may benefit the host alga rather than damaging it.

Topshells (trochids)

The four common species of topshells in north-west Europe show varying distributions. *Gibbula cineraria* and *Calliostoma zizyphinum* occur from Spain to Norway, while *Gibbula umbilicalis* and *Monodonta lineata* are found on the west coasts of Britain and Ireland, and further south, but not in the north. *Gibbula pennanti* is found in the Channel Islands, and is common on the Brittany coast.

Growth rates

Although *Monodonta lineata* is at its northern limit in south-west Britain, many populations there receive regular recruitment each year (Kendall, 1988). In this species a growth check (a line around the mouth of the shell) forms each winter, and from the number of checks one can place any individual into its year-group or 'cohort'. Snails 2–3 mm in diameter have no growth checks, while those between 3 and 14 mm have one check. Above this size, the growth rate slows down, and 7-year-old snails reach a maximum width of about 25 mm.

Diurnal migrations of *Gibbula cineraria*

Gibbula cineraria, which lives primarily in the shallow sublittoral zone, shows quite different distributions in the day and at night (Fig. 5.9). In Lough Hyne, Ireland, numbers visible on the tops of rocks increase rapidly at dawn and decrease rapidly at dusk precisely. This response is unaffected by state of tide, but is restricted to larger individuals—the small *Gibbula* seldom, if ever, migrate.

Fig. 5.9   Diurnal migrations of the topshell *Gibbula cineraria* on the sea bed of Lough Hyne, Ireland. Open and closed circles show results in two separate areas of 1 m². Snails move up to the tops of rocks (and therefore become visible) in daytime and retreat in darkness, and this behaviour is independent of tides. (After Thain, 1971.)

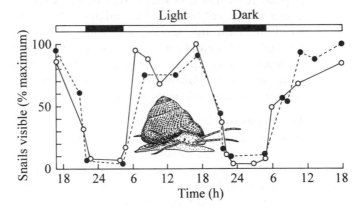

Is the response of *Gibbula* to light and dark a direct response to illumination, or is it influenced by other conditions or by some form of activity rhythm? Field experiments were carried out by Thain (1971) in which an area of the shallow sublittoral rocky bottom was made completely dark at various times of day. The area to be darkened was covered with a bottomless box through which sea water was pumped continuously. The box had little effect on *Gibbula's* movements, however, suggesting that the migrations are governed by an endogenous rhythm and not by a direct response to light.

The timing of this rhythm can be altered experimentally, but only to a small degree. When artificial light was turned on over an area of sea bed, in the middle of the night, a few *Gibbula* came up, but soon went down while they were still illuminated. When illumination was turned on shortly before dawn, upward migration occurred earlier than in the control square. Thus upward migration can only be induced near the correct time.

Control of zonation in *Gibbula umbilicalis*

Gibbula umbilicalis is found in the middle and lower half of the littoral region. Individuals displaced into the shallow sublittoral region move directly towards their proper intertidal level. How is this reaction mediated? Thain *et al.* (1985) collected *G. umbilicalis* and *G. cineraria* from one shore, marked them, and moved them to the shallow sublittoral region of a different shore. Over the next few days, *G. cineraria* scattered in all directions whereas *G. umbilicalis* moved only towards the land. It did this even though collected from a south-facing shore and now finding its way up a north-facing shore, so compass orientation, sun navigation and topographic memory are ruled out as orientating mechanisms. However, a visual stimulus must be involved because in further experiments the topshells moved towards a black plastic wall which was erected offshore. Exactly how this visual response works has yet to be determined.

Mesograzers—amphipods and isopods

The molluscan species so far considered are all large enough to be termed macrograzers. They are prominent on rocky shores, and for many years research has concentrated on them in preference to smaller forms. There are, however, many other grazers, including small gastropods, polychaetes, and particularly amphipods and isopods. These are inconspicuous, and although they occur in large numbers they have attracted little attention. Nevertheless, these 'mesograzers' may have significant effects on rocky shore ecosystems (Brawley, 1992).

Gammarid amphipods

There is an immense variety of gammarid amphipods in the intertidal zone and the shallow sublittoral. Many of these animals are associated with algae, and in particular we shall discuss *Hyale* spp., on which studies have centred.

On the shores of New England and of north-west Europe, *Hyale nilssoni* is the most common alga-inhabiting amphipod. It is abundant in fucoids such as *Fucus spiralis* and *Pelvetia*, but prefers the red epiphyte *Polysiphonia lanosa*. In the laboratory *Hyale* is attracted to *Polysiphonia* more than to fucoids, and consumes more of it. Since many algae produce chemicals that deter grazers, it may be that *Polysiphonia* has less chemical defence than the fucoids. However, physical structure and toughness are also important (p. 54).

On the north coast of Spain, *Hyale nilssoni* is an important grazer of *Fucus vesiculosus* (Viejo and Arrontes, 1992). It feeds on the fucoid for most of the year, eating mainly the damaged parts of the thallus. Another grazer here is the isopod *Dynamene bidentata*, and it appears that the feeding rate of *Hyale* is *increased* when *Dynamene* is present. *Hyale* can probably feed on the areas already damaged by the isopod, so that instead of competing, the isopod facilitates feeding by the amphipod.

On the coast of Chile, *Hyale hirtipalma* has quite a specialized relationship with red algae of the genus *Iridaea*: *H. hirtipalma* preferentially eats the mature reproductive tissue, and the spores survive passage through its gut. The number of spores settling successfully is *increased* by this, so that the amphipod actively *aids* dispersal by the alga.

Epiphytic feeders: do amphipods aid or hinder macrophytes?

Many amphipods eat epiphytes rather than the macroalgae on which they grow. Epiphytes decrease light reaching the host, and may increase wave damage and dislodgement by increasing drag, so there has been much discussion about whether amphipods which eat them can increase the viability and growth rates of macroalgae. However, experiments on the amphipod fauna of the most common seaweed in North Carolina, *Sargassum filipendula* (Duffy, 1990), suggest that while amphipods such as *Caprella pennantis* can reduce epiphyte growth significantly they do not promote the growth of the host (Fig. 5.10). Other species of amphipod actually eat the *Sargassum*, in which case it decreases in mass.

It is possible that very large numbers of amphipods could be beneficial to macroalgae, but that they are usually kept below their carrying capacity by predators such as fish. There have as yet been too few studies to be able to say how common this might be.

Idotea in the Baltic

Intertidal fucoids usually harbour numbers of isopods such as *Idotea* spp. Normally, *Idotea* probably eats a mixture of epiphytes and the old eroding *Fucus* thalli, and food resource and grazer maintain a fluctuating equilibrium. In the Baltic, where species diversity is low, *Fucus vesiculosus* is the major coastal alga and isopods such as *Idotea baltica* are the major herbivores. In the late 1970s, *Fucus vesiculosus* suffered extensive decline, and there have been several attempts at interpreting the underlying causes (Vogt and Schramm, 1991). A complex sequence of interactions seems to have occurred, triggered at least partly by rising nutrient concentrations. Eutrophic conditions allowed filamentous epiphytes such as *Pilayella littoralis* to grow enormously. The increased epiphyte load had two effects. First, it allowed survival of large numbers of juvenile *Idotea*, which feed mainly on filamentous algae. Secondly, it led to a decline of *Fucus* plants through reduced settling of sporelings and reduced reproductive capacity. The *Fucus* belt thus aged gradually without addition of young plants. Since adult *Idotea* feed more on decaying macroalgae than on epiphytes or growing fronds, this condition provided an ideal environment for growing numbers of isopods. In some areas, the original ageing *Fucus* stands were entirely destroyed by grazing, although later settlement in the 1980s replaced the algal cover.

Laboratory experiments demonstrate the potential effect of grazing by *Idotea*. Adult *Idotea baltica* prefer the oldest stipes of *Fucus vesiculosus* when given a choice of various segments of the plant. Individuals can ingest more than 25% of their body weight of *Fucus* each day, and when allowed to browse on whole plants can reduce them to skeletal fragments.

Fig. 5.10   The relations between growth of *Sargassum filipendula* and its algal and amphipod epiphytes. Experiments were carried out in outdoor tanks, starting with epiphyte-free plants. In treatment 1 (no amphipods present), *Sargassum* grew well, as did its algal epiphytes. In treatment 2 (the amphipod *Ampithoe* added), *Sargassum* growth was apparently decreased (though not significantly so), and epiphytes decreased: *Ampithoe* eats *Sargassum*. In treatment 3 (the amphipod *Caprella* added), *Sargassum* growth was high, but there were few epiphytes: *Caprella* eats epiphytes. Bars show SE. (After Duffy, 1990.)

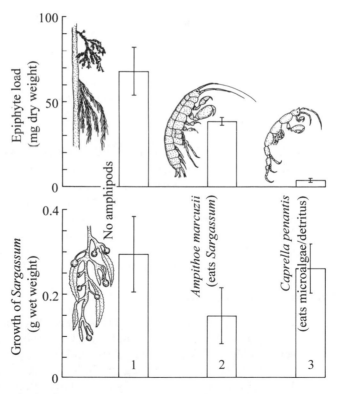

The *normal* effect of mesograzers on *Fucus vesiculosus* populations is probably minimal, because the ageing plants, when eaten, are replaced by newly settled thalli, which are unpalatable. Only when the normal succession of algal replacement is broken does grazing on older plants produce a striking consequence for the community.

Sea urchins

In the shallow sublittoral around the world, sea urchins graze on algae, on the invertebrates associated with or attached to these algae, and on dead plant material. In north-west Europe *Echinus esculentus* is the largest urchin at the bottom of the shore. *Paracentrotus lividus*, a warm-water species, is found in tide-pools along the west coast of Ireland, in Brittany and the Mediterranean. In North America several species of urchins are common and *Strongylocentrotus droebachiensis* may play an important part in limiting distribution of kelp on the east coast. On the Pacific coast, *S. franciscanus* and *S. purpuratus* are common.

Settlement and growth

The eggs and larvae of sea urchins are pelagic, so it is important that larvae should settle in the 'correct' microhabitat. Chemical cues are probably important here, since extracts of coralline algae encourage settlement of *Strongylocentrotus purpuratus* which lives on corallines and on algal turf (Rowley, 1989). The time immediately after settlement is probably a hazardous one for urchins. In southern California the very small juveniles of *Strongylocentrotus franciscanus* shelter under the spine canopy of their own species, and so are protected from predators while small. At a diameter of 30–40 mm, however, they leave the protective spine canopy and are susceptible to predators until they grow too large to be attacked.

The growth of urchins can be studied by examining the 'growth rings' laid down in their calcium carbonate plates. Laboratory experiments in France show that *Paracentrotus lividus* can increase its diameter by 0.3–1.9 mm per month, depending upon food supply. Some urchins, however, are actually able to shrink if food supply declines. For instance, the Pacific *Diadema antillarum* can decrease the size of its test while keeping the feeding apparatus almost the same size, so that examination of each individual can give an indication of its past feeding history.

The *Paracentrotus* community

The urchin *Paracentrotus lividus* has a great effect upon the sublittoral community. Although we consider in detail the factors that structure communities in Chapter 8, it is worth while taking a brief preview here of the influences that one grazing species can have (Kitching and Thain, 1983). In areas of Lough Hyne, Ireland, that are dominated by *Paracentrotus*, there is no noticeable growth of any algae other than corallines. Outside the grazing areas, however, the shallow sublittoral is covered by soft upstanding algae, such as *Stilophora rhizodes* and *Codium fragile*. When part of a bay dominated by *Paracentrotus* was completely cleared of urchins, *Enteromorpha* grew over the cleared area and reached a coverage of 100% within a year. In areas where *Paracentrotus* remained, it grazed the soft upstanding algae and thus permitted the growth of crustose coralline algae, especially *Lithophyllum incrustans*, which is resistant to grazing. This in turn provided space for limpets (*Patella aspera*) and for the bivalve *Anomia ephippium*, which is stuck down to the crustose surface. The corallines also accumulated many small tube-dwelling polychaetes which burrowed within them, so that the whole structure of the community was different in grazed areas and non-grazed areas.

What controls sea urchin densities?

In Europe, *Paracentrotus* is attacked by crabs which may control population size. Since urchin gonads are a prize food for humans, predation by man can also decimate populations. In North America, *Strongylocentrotus* species are to some extent controlled by their predators. On the Pacific coast of North America, one of the major predators of urchins in certain areas is the sea otter, *Enhydra lutris* (Van Blaricom and Estes, 1988). There has been much discussion over the possibility that the otters act as 'keystone predators', controlling urchin populations, and therefore determining community composition over wide areas. This is such an important concept that we shall defer it for discussion in more detail (see p. 183).

Other possible predators are lobsters and fish. In Canada, lobsters were offered a choice of food in underwater cages, and of six prey species they preferred first the crab *Carcinus maenas* and then equally *Cancer irroratus* and sea urchins. Later experiments confirmed that lobsters can have a significant impact on urchin populations (Hagen and Mann, 1992)

On Santa Catalina Island, California, the sea urchin *Centrostephanus coronatus* is abundant at subtidal sites dominated by giant kelp, *Macrocystis pyrifera*. The urchins remain inactive in holes in the rock during the day, but emerge soon after nightfall to forage up to 1 m away. They come back to the same hole considerably before daybreak. By this behaviour urchins avoid meeting the sheephead wrasse, *Pimelometopon pulchrum*, which forages by day (Fig. 5.11). If an urchin is removed from its hole and put out by day, as though to forage, it is quickly and vigorously attacked by sheepheads. The urchin is turned over so that its less well-defended oral side faces upwards, and this is broken in. A frenzy of feeding by many sheepheads and other predatory fish follows and the entire contents of the urchin are devoured in less than a minute. Here a predator has strongly influenced the evolution of urchin behaviour, but there is little evidence that it gets the chance to eat many urchins at the present time.

Large predators are not the only organisms to attack urchins. Enormous numbers of the sea urchin *Strongylocentrotus droebachiensis* on the Atlantic shore of Nova Scotia died in the autumns of 1980–83 from a disease caused by a minute amoeboid organism, *Paramoeba* sp. (Miller, 1985). Urchins suffering from the disease are unable to turn upright if placed upside down, cannot extend their tube feet fully, and allow their teeth to gape. Their spines droop and eventually fall off, and death follows.

Fig. 5.11   Movement of the sea urchin *Centrostephanus coronatus* out of its shelters in holes and crevices, at Santa Catalina, California. Dotted lines enclose the extreme values. The urchins are entirely nocturnal, and thereby avoid predation by the sheephead wrasse, *Pimelometopon pulchrum*, whose activity period is indicated by the solid curve. (After Nelson and Vance, 1979.)

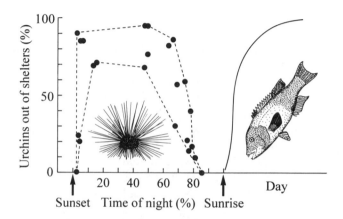

While the direct influence of predators and disease may in many cases control urchin abundance, there is also evidence that they may be controlled by food supply. In California, urchins become destructive grazers during some El Niños (the spread of warm water currents across the Pacific), but in others the supply of algae appears to be so reduced that urchin populations remain low. Since grazing by amphipods and other species may aggravate the low algal supply, it is not surprising that the grazing activities of many species have overlapping effects.

Experiments to investigate grazers

The grazers on rocky shores can best be investigated by a combination of field and laboratory studies, and, in principle, by combining long-term and short-term observations.

Distribution of limpets

How does size, shape and density of limpets (nos/m^2) vary with exposure? Length, width and height of shells can be measured without detaching limpets. A *few* should be detached to ascertain species. If carefully done, these should reattach.

On a smaller scale, how does limpet distribution relate to distribution of macroalgae? Are limpets clumped, evenly dispersed, or spaced at random? Do they change their home scars over short/long periods? This can be investigated by making careful maps of limpet groups, or, better, by painting reference marks on each limpet and on the rock nearby.

Limpet responses to predators

Test the reactions of limpets to the presence of predators such as the starfish *Asterias*. This can be done in rock pools, by placing the predators near or on top of the limpets. Alternatively, limpets on mobile rocks can be placed in an aquarium. Can you observe stomping, mushrooming or escape? Note that limpets detached from their home scar will *not* respond. Non-predatory starfish such as *Asterina* can also be used, and the responses of topshells such as *Monodonta* and *Gibbula* can be assessed.

Population structure of *Monodonta*

Using growth check lines, *Monodonta lineata* can be grouped successfully into year classes. Plot size–frequency histograms, and compare population structures on different shores or at different vertical heights.

Response of grazers to chemical cues

Begin by recording the distribution of topshells or winkles in relation to distribution of macroalgae. Then test reactions of snails to the various algal species. This can be done using either whole fronds, or extracts made by crushing or mincing the algae in sea water.

Distribution and behaviour of mesograzers

Most amphipods are hard to identify, but *Hyale nilssoni* is usually characteristic of the high-shore *Pelvetia* zone. Once identified there, following its distribution is quite easy. Is it restricted to *Pelvetia*? Does it respond positively to the plant or its extracts in the laboratory? Is its distribution the same on exposed and sheltered shores?

Grazing mechanisms in a variety of gastropods

The variety of radula structures gives a good indication of how snails feed. This can be supplemented with examination of gut contents, faecal pellets, and the surfaces on which grazing has occurred.

Long-term exclusion experiments

If you have access to the same shore at regular intervals, it is possible to carry out experiments in which grazers are excluded from set areas. Limpets can be removed from *small* areas (1 m^2) by hand, and continued exclusion can be maintained by regular (weekly) visits, by surrounding the exclusion area with a strip of antifouling paint, or by erecting mesh fences of galvanized or plastic-coated wire. Fencing is a time-consuming process, however, and needs a battery-operated drill. Do NOT attempt to use a mains drill and generator on the shore. Exclusion areas, with suitable controls, should be monitored at least monthly for maximum information.

6 Suspension feeders: how to live on floating food

Sessile animals outnumber the mobile ones enormously on rocky shores: on the tops of rocks, barnacles and mussels may entirely cover the rocks themselves, while on the undersides and on many macroalgae, an obscuring film is formed by bryozoan colonies, ascidians, sponges, hydroids and polychaetes. All these species depend upon food brought to them in suspension, but many have also developed methods of aiding or channelling the local flow of water to increase their efficiency of particle capture.

When the adults are sessile, the behaviour of pelagic larvae is of extreme importance in determining their distribution, and many larvae have complex specializations allowing them to settle in appropriate micro-habitats. A mixture of sessile species also settles on man-made structures, and the understanding of the ecology of these so-called 'fouling communities' is therefore of great importance when it comes to attempts to clean the hulls of ships or the cooling-water pipes of power stations.

The distribution of suspension feeders

Natural substrates

On wave-exposed shores throughout the world, barnacles and bivalves are usually the dominant suspension feeders. The barnacles are for the most part sessile or 'acorn' barnacles—that is, they are covered by a set of rigid calcareous plates in the shape of a volcano, cemented directly to the substrate. There are many genera, and the individuals vary in size from an average of perhaps 5–10 mm in diameter, up to the giant forms of the Pacific coast of America, which reach 10 times this width. Also on the Pacific coast of America, acorn barnacles are accompanied by stalked barnacles such as *Pollicipes*—those with a flexible fleshy peduncle anchoring the body to the rock (Fig. 6.1).

The bivalves of exposed shores are primarily mussels—smooth-shelled and dark in colour, and attached to the rock by a bundle of fibres called the byssus. Two genera, *Mytilus* and *Modiolus*, are common world-wide and often form dense, single-species zones. In warmer areas of the world, there may also be oysters on the shore—rough-shelled bivalves cemented to the rock by one valve. Oysters are common in some parts of American and Australasian coasts.

Besides the barnacles and bivalves, there are many other abundant suspension feeders in different parts of the world (Fig. 6.1). At the bottom of some exposed shores in Australia, southern Africa and South America, the dominant animals are large ascidians of the genus *Pyura*. Individuals of these tunicates may be up to 15 cm high, and their rough, leathery tunics form a striking zone. In other areas, the place of barnacles and mussels may be taken by suspension-feeding polychaetes. Some of the Serpulidae, which form calcareous tubes, may form thick layers on the rock: in Australia, *Galeolaria* is widespread, and in southern Africa *Pomatoleios* grows into dense masses. *Pomatoceros* is common in Europe, but does not form masses. Where rocky shores are near a good supply of sand, different species of gregarious polychaetes may form coatings of sandy tubes. In Europe and the Atlantic coast of North America, *Sabellaria* forms sizeable reefs, and *Gunnarea* has a similar form in southern Africa.

Fig. 6.1   The occurrence of some suspension feeders. Note that mussels and barnacles occur world-wide.

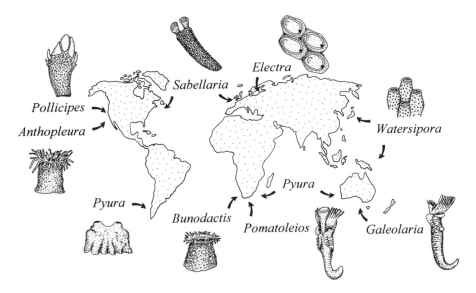

Other groups on rocky shores that feed primarily on plankton include the sea anemones, although many of these feed on larger animal prey using nematocysts. Genera such as *Actinia* are common in Europe and around the world. In southern Africa, *Bunodactis* may form sheets of closely packed individuals, covering the rock. On the Pacific coast of North America, *Anthopleura* may do the same.

The remaining groups of suspension feeders are very often most common not on the rock surface, but under overhangs, on the lower side of boulders, or growing on seaweeds. As a result they are often overlooked, but in fact the bryozoans, sponges, ascidians and hydroids can be exceedingly abundant, and often completely obscure the surface of the substrate to which they are attached. In some areas the bryozoans that have calcareous skeletons can form reef-like masses—in New Zealand, Australia and Japan *Watersipora* is an important fouling organism on ships and on rocks, while *Electra crustulenta* forms substantial low-shore and subtidal reefs in the Baltic. Colonial ascidians, unlike the large solitary *Pyura*, may also grow as a thin layer covering rocks and algae, alternating with sponge colonies. Although the individuals of all these colonial forms are small, the overall filtering capacity of colonies is immense.

Man-made substrates, 'fouling' and its prevention

Many man-made structures provide extensions of the hard surfaces naturally found on rocky shores. The hulls of ships, the tubes of cooling systems, the surfaces of docks, piers, buoys and offshore platforms all, therefore, attract diverse communities of sessile fauna and flora. These have some parallels with 'weeds' on land, and are known in general as the 'fouling community'.

Fouling communities consist of a mixture of algae and suspension-feeding invertebrates, although most research has concentrated upon the invertebrate component. Mussels, in particular, are the cause of blockage in cooling systems. Fouling on ships increases drag, thereby reducing speed, and growth of barnacles may increase the possibility of corrosion. Growth on offshore buoys, particularly by algae, again increases drag and may eventually result in the buoy dipping below the water surface in strong currents.

The development of methods for preventing or controlling fouling has not been easy. The methods must be safe for personnel using them and safe for the environment, yet be effective in deterring settlement and growth of fauna and flora. They must not have adverse effects on the system being protected, and must function in a variety of conditions.

In the cooling-water systems of power stations, use of copper–nickel tubing deters fouling communities effectively, and treatment of cooling water with chlorine reduces the residual problem to manageable levels. In general, other situations employ techniques that rely upon the application of antifouling coatings, combined with periodic mechanical cleaning. Many substances have been employed, ranging from copper compounds to organic–metal complexes. Some of these, such as tributyl-tin (TBT) are exceedingly toxic and are effective as antifouling treatments. Unfortunately, as we shall see when discussing predators on rocky shores (p. 144), TBT also has long-term effects on organisms that are *not* part of the fouling community.

The great practical importance of marine fouling has led to an enormous volume of work on fouling organisms. If more were known about how larvae of these species go about settling and growing, it might be possible to devise more effective ways of deterring them. The practical problem has been the driving force behind an attempt to understand the ecology of fouling, at least in terms of larval settlement.

The problems of sitting still: how to be sessile

Most of the suspension feeders on rocky shores are sessile, though this is not true on soft shores. Being sessile means being dependent upon local conditions for a supply of food and nutrients. It also means that organisms must tolerate the conditions of flow, wave action and desiccation, rather than being able to move away from them like the mobile grazers and predators. Many suspension feeders are also small, and the details of water flow near to the substrate are critical factors for them. For all these reasons, the relationships between suspension feeders and water currents are particularly important, and we consider these direct hydrodynamic problems first, before moving on to the indirect effects that relate to food and feeding strategies.

Water flow near the substrate

The way in which water flows over a rocky surface depends upon a number of factors, including the density and viscosity of the water and the shape and roughness of the rock. The dominant factor, however, is velocity of flow. At slow flows, the water tends to move smoothly over the rock, and can be thought of as consisting of a number of non-mixing layers. This is called laminar flow. Each hypothetical 'layer' experiences friction with those above it and below it, which slows down the flow, and the layer nearest to the substrate does not slip relative to it. There is therefore a velocity gradient from the substrate (where velocity is theoretically zero) upwards to the 'mainstream' velocity, above which it is constant. The region of this gradient in velocity is called the 'boundary layer' and the dimensions of this layer are vitally important for suspension feeders. If they are small enough to live within it, they will not experience the drag forces of the mainstream current—but, as a corollary, local current speeds may not be high enough to produce an adequate food or oxygen supply. If, on the other hand, they project into the mainstream flow, they will receive maximum oxygen and nutrients, but they must be able to tolerate or resist the large effects of current drag found there (Fig. 6.2).

Fig. 6.2   An example of the boundary layer in turbulent flow. Within about 1 cm of the substrate, water velocity is slow, and the barnacle *Semibalanus balanoides* can feed in a current of less than 0.2 m s⁻¹. Above this height, however, velocity rises logarithmically to the 'mainstream velocity', in this case 1.0 m s⁻¹. The straps of the alga *Himanthalia elongata*, which may reach 2 m in length, project into this stronger flow. (After Denny, 1988.)

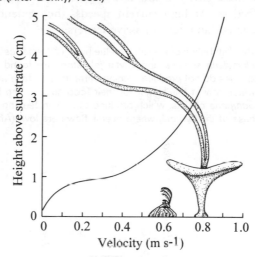

The situation is more complicated at higher water velocities because the 'layers' then break down and water flow in the boundary layer becomes turbulent. Layers mix with those above and below, and the supply of nutrients and particles to the region immediately above the substrate increases. Drag forces, however, tend to become variable both in size and direction and this may make them more difficult to deal with. These effects are further emphasized in waves, where currents reverse rapidly, and accelerational forces become of more importance than the drag that is produced by steady flow (p. 36).

Feeding in relation to water flow

Suspension feeders live in an environment that has low concentrations of food, in the form of small particles such as plankton and detritus. There is a plethora of mechanisms for obtaining this food, ranging from so-called passive methods such as the deployment of a filter in the mainstream flow, to active mechanisms in which animals create currents past their feeding surfaces. Some animals can feed both actively and passively.

Whether animals use active or passive mechanisms for suspension feeding relates partly to their position in or above the boundary layer. For example, of the epizoites on the colonial hydroid *Nemertesia*, the more passive feeders such as hydroids are found on the distal regions (in or near the mainstream current), while active feeders such as bryozoans, ascidians, bivalves and sponges are found near the hydroid's base, i.e. well inside the boundary layer (Fig. 6.3). Changes in behaviour can also take advantage of variation in the flow. Many species of barnacle show extremely variable behaviour, depending upon current speed. Some balanoids, for example, generate a rhythmic beat with their limbs ('cirri') at low flows, when they would be classified as 'active' feeders. At high current speeds, they extend the cirral net into the current and act as passive feeders.

Fig. 6.3 The distribution of epizoites on the hydroid *Nemertesia antennina*. Hydroids such as *Plumularia setacea* and *Clytia johnstoni* are found near the distal ends. These species do not produce feeding currents, and take advantage of the high external water flow at this height for their food supply. The bryozoans *Electra pilosa* and *Scruparia chelata*, which produce their own feeding currents, are found near the base of the hydroid, where overall flows are low. (After Hughes, 1975.)

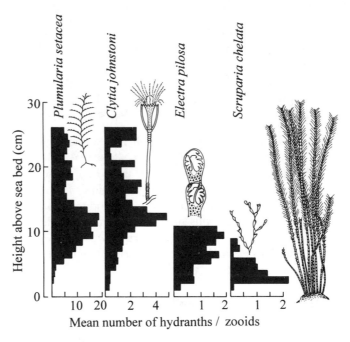

The variety of suspension-feeding mechanisms

An enormous variety of structures can be used to catch suspended food. Most devices depend either upon some kind of appendages covered in hairs or 'setae' (these are common particularly in the Crustacea) or upon mechanisms involving mucus, mucus-with-cilia or cilia alone (characteristic of soft-bodied invertebrates such as sponges, bivalve molluscs, ascidians, bryozoans and polychaetes). In addition, coelenterates use nematocysts to trap prey.

The structure of these various mechanisms is well known, but only recently have the details of how particles are actually separated from sea water emerged. There are three basic mechanisms (LaBarbera, 1984).

One of these is the so-called 'scan-and-trap' method: a steady stream of water is moved past the animal until a particle approaches. This is detected, and the flow of water is then altered to capture the particle and its water shell. One of the major unknowns here is how particle detection occurs. A second method is mechanical sieving, which, in spite of the apparent resemblance of many feeding structures to sieves, is rare. In this method, water is forced through a filter and larger particles than the mesh size are retained, while *no* particles smaller than the mesh are retained. This is a costly process because it is difficult to force water through a mesh of 1–10 μm: viscosity becomes a dominant force at these dimensions. Nevertheless, some of the cilia of bivalves (the 'laterofrontals') form a mesh, and the cirri of barnacles may do the same. The third, and most commonly found, method is called 'aerosol trapping'. Here the water is forced through a mesh, but the filtering elements have some adhesive property—either they are covered in mucus, or their cilia can redirect particles. Particles much smaller than the mesh size therefore stick to the filtering elements and are then conveyed to the mouth. Some variants of this method occur in bryozoans, ascidians and sponges, while hydroids use their nematocysts to capture particles that land on their tentacles. Note that more than one of these mechanisms may be used, depending upon external conditions.

The majority of suspended food caught on rocky shores is probably phytoplankton. Since this can range down to sizes of the order of 1 μm, the capturing mechanisms have to be exceptionally fine. They are also remarkably selective, and do not seem to depend solely upon size of particle. A good example of selectivity is presented by the bivalve *Ostrea edulis*. When this species was supplied with a mixture of algal food—a diatom, a dinoflagellate and a flagellate, all of similar dimensions of 10–20 μm long—it showed *at least* three selective mechanisms. The gills, the gut and organs near the mouth called 'labial palps' can each differentiate between the algal species. *Ostrea* is therefore very far from being an unselective suspension feeder. From what we know of other suspension feeders, the same is likely to be true of the majority, even though the mechanisms of feeding are not themselves clear in all cases.

The food supply of suspension feeders

Food available in suspension is enormously varied in origin, and also shows a great deal of variation over time and from one locality to another (Okamura, 1990). Many suspension feeders take up primarily phytoplankton, which is most abundant overall at the time of the spring bloom, but which is concentrated in patches rather than having an even distribution. The sizes of phytoplankton vary from a few micrometres to over 100 μm.

Zooplankton is less abundant than the phytoplankton on which it feeds, but may show even larger fluctuations in abundance than the phyto-plankton at any one place because many species undergo vertical migration. There are also species that become planktonic only at night, as well as those that occur in the plankton only as larvae.

Even larger animals provide the diet of some of the coelenterates. Some sea anemones, for instance, feed mainly by ingesting mussels and barnacles dislodged from the rocks by waves, or on prawns and other mobile crustaceans that are trapped by nematocysts as they pass by. Whether these should strictly be considered suspension feeders is a point for discussion, but since they consume food that is 'brought to them', we will include them here.

Besides the living matter listed so far, suspension feeders can tap various 'dead' sources of energy. Particles of organic detritus are abundant in some areas, and these may have their energy value enhanced by colonies of bacteria, fungi and protozoans. In addition, most soft-bodied marine invertebrates can take up dissolved organic matter directly through the epidermis. Although this mode of nutrition is thought to be of minor importance in most species, some of the deep-water gutless animals depend upon it to a large extent.

Finally, some suspension feeders are associated with symbiotic micro-algae or bacteria, which fix carbon and transfer some of it to their hosts. Corals and giant clams are well-known examples, but some sea anemones contain symbiotic dinoflagellates, and many sponges have bacterial and cyanobacterial symbionts.

The effects of suspension feeders on plankton populations

If sessile invertebrates depend upon plankton for their food supply, an important question to ask is whether their feeding processes have any significant effect on plankton populations (Bayne and Hawkins, 1992). There are three main ways in which suspension feeders could affect the planktonic system. First, they could deplete or control phytoplankton biomass. Secondly, they could affect the amount of dead particulate material in suspension. And thirdly, they could affect levels of dissolved nutrients.

The suggestion that suspension feeders might control phytoplankton biomass has been considered many times by using 'budgeting' studies for particular systems. Unfortunately, these have had to depend upon many assumptions and extrapolations and it has been impossible to make definite conclusions. Field studies using large enclosures (often called mesocosms) have been more successful. In a Danish fjord, for instance, the addition of strings of mussels (*Mytilus edulis*) to cylindrical plastic enclosures decreased the phytoplankton levels by up to 60% of those in the controls. The mussels had most effect in September, probably because at this time most of the phytoplankton consisted of relatively large cells, easily filterable. In the spring and summer, most of the plant cells were only 1–2 μm in diameter, and mussels are not efficient at filtering particles of this size.

Mesocosm experiments do not, however, provide proof that suspension feeders have any significant effect in the wild. There have been various suggestions that, at least in estuarine systems, where the tide advances over wide expanses of flats, the low-shore suspension feeders might partially clear the incoming water of food, so that those animals living higher up the shore have less to eat. Such studies, however, suffer from the problem of confounded variables—there are too many effects that might cause the phenomena seen, to be able to establish cause and effect.

Two reports from mussel beds do, however, demonstrate the filtering power of *Mytilus edulis*. In Killary Harbour, on the west coast of Ireland, where concentrations of chlorophyll could be measured upstream and downstream of a commercial mussel raft, mussels cleared 47% of chlorophyll *a*. In the St. Lawrence Estuary, Québec, mussels reduced phytoplankton levels immediately above them to about half the concentration in surface water. Experimental studies elsewhere have suggested that other suspension-feeding communities are also able to make significant reductions in plankton concentration.

Suspension feeders may also affect suspended material and nutrient levels offshore. Material that is not passed to the mouth by the labial palps of bivalves is formed into 'pseudofaeces', and ejected. Pseudofaeces can form a very high proportion of the material filtered at the gills, so that large deposits can be formed by beds of molluscs. These deposits can then be resuspended, adding to available particulate matter. The excretion of nutrients by bivalves, especially ammonia, may also stimulate plant growth if phytoplankton is nutrient-limited. Overall, the effects of suspension feeders on coastal ecosystems may be much greater than previously suspected.

Mussels

Mussels dominate in much of the low and mid intertidal region in temperate seas of the northern and southern hemispheres. There are many genera, all with a narrow anterior end and with the anterior adductor muscle reduced. The pointed 'umbo' is at the anterior end of the shell in *Mytilus*, and slightly set back in other genera, including *Modiolus*.

The taxonomy of mussels in north-west Europe is not entirely clear. Shell shape is used to distinguish *Mytilus edulis* from *M. galloprovincialis*, but shape has now been shown to be greatly affected by environment. When

Mytilus edulis was transferred from a wave-exposed site in Yorkshire, England, to a sheltered site, it developed a relatively greater increase in height than in length, compared with controls, so that it appeared more like *M. galloprovincialis*. Furthermore, the distributions of *Mytilus edulis* and *M. galloprovincialis* are sometimes quite distinct, but in other places hybridization may take place (Seed, 1974), so that intermediates occur. In view of the difficulties of separating the two species, there remains considerable doubt as to the validity of the distinction between them.

Mussel patches and factors affecting distribution

Mytilus edulis is essentially an intertidal organism, which can form extensive beds dominating the rock surface. It can also form strips or patches. At wave-exposed sites in Yorkshire, England, there are mosaics of *Mytilus* and *Semibalanus* on the upper shore, continuous beds of relatively large mussels on the mid shore, and mosaics of small mussels on the lower shore. Communities such as these undergo considerable changes over time due to the effects of factors such as disturbance and competition.

A good example of this is shown by a study near St. Malo, Brittany, where mussels underwent massive fluctuations over a 9 year period. Initially, they were fairly abundant, mainly from MTL to MHWN. Barnacles (*Semibalanus balanoides* and *Chthamalus*) covered the rock and all the mussels present, and dog whelks (*Nucella*) were abundant. There was an extensive spat-fall of mussels in the fourth and fifth years of the study, which in turn covered the barnacles. The dog whelks preyed mainly upon barnacles for a further 2 years, but by then there were none left. Then they changed their diet to mussels, and began to multiply greatly. The dog whelks preferentially bored the mussels at the bottom of the mussel bed, against the rock, and this led to the tearing off of patches of mussels. In the ninth year, barnacles reoccupied the areas stripped of mussels, and the dog whelks returned to feeding on barnacles.

On the Pacific coast of North America, competition between the two mussel species is an important factor govering distribution (Bayne, 1976). *Mytilus californianus* extends from Alaska to Mexico and requires a very wave-beaten coast, where it dominates from above MTL to below the lowest tide. It has a ribbed shell, much heavier than that of *Mytilus edulis*, and forms much stronger byssus threads, so that it is better protected against predators and against being torn off by rough seas. *Mytilus edulis* dominates at sheltered sites, where it can detach its byssus threads and move to the surface of the mussel bed or overlying ground, but on the open coast it is an opportunist. When storms rip off patches of *M. californianus*, the bare patches are quickly colonized by *M. edulis*, but although the young of these grow faster than young *M. californianus*, they are eventually outcompeted.

Differences in resistance to physical factors and predation also affect the distribution of these two mussels. *Mytilus edulis* can live higher on the shore than *M. californianus* because it is more tolerant of both desiccation and frost. *Mytilus californianus* is more resistant to drilling by the gastropod *Thais* and can grow too big for the starfish *Pisaster ochraceus* to open it, so that it can live lower on the shore. Thus the two species exist on the same shore, but occupy different niches.

Larval biology

The first larval stage of *Mytilus edulis* is the ciliated trochophore, which lasts only for about a day. It is followed by various stages of 'veliger', lasting in all about a month. This has a pair of shell valves, and carries a ciliated swimming organ or 'velum'. It settles first on filamentous seaweeds such as *Polysiphonia lanosa* (an epiphyte of *Ascophyllum nodosum*) and *Ceramium rubrum*. Maximum first settlement in north Wales occurs in June–July, but in July–August these 'early plantigrades' detach themselves and are carried once again by water currents. 'Late plantigrades' settle in their final position on the mussel beds in July and August (Fig. 6.4).

Fig. 6.4 Settlement of *Mytilus edulis*. In 1966/7, pediveligers in the plankton near Whitby, England, peaked in July (numbers per 20 min plankton tow). Early plantigrades settled on the alga *Polysiphonia*, the peak being in August (numbers/g dry wt of alga). These plantigrades detached and late plantigrades then re-settled on a fibre-glass panel (numbers/ 10 cm^2), peaking in October. (After Seed, 1969.)

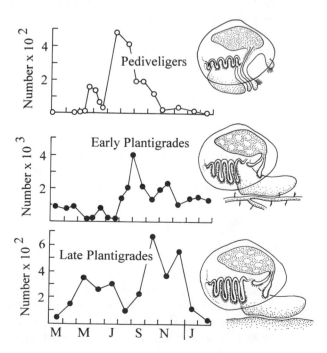

Spawning and settlement dates vary from place to place, however. In Ireland, there is usually extensive spawning from March to September, and settlement may therefore occur over much of the year. Laboratory studies on the larvae of *M. edulis* have shown that growth is much influenced by food supply (Pechenik *et al.*, 1990), and this could influence the variation in settlement. Metamorphosis can also be delayed for at least 45 days, if appropriate types of substrate are not present.

Growth

Growth rate varies greatly and depends largely on the amount of time available to the mussel for feeding. In experiments in New Brunswick, *Mytilus edulis* grew in inverse proportion to emersion time, and those emersed for 75% of the time hardly grew at all. The position within the mussel bed is also of importance: individuals in the centre of a bed grow slowly, while those at the edges grow faster but may pay the penalty of suffering more predation by crabs.

It is not, in fact, easy to measure growth on the shore, but three methods can be used. The first depends upon reproduction being limited to one season in the year. In this case size-frequency analysis of the population shows a series of peaks, each representing a year. Growth can be followed for 2–3 years by following these peaks, but after this time they join up, no doubt because of variations in the growth rate. Studies of the growth of mussels in experimental growth boxes on the shore have therefore been more helpful. They have shown growth rates to be much the same as those deduced from a third method, in which 'disturbance rings' on the shell are counted. Many marine organisms show some kind of 'disturbance ring' (see pp. 65, 95, 100), such as the rings seen in fish scales, but these are not always laid down annually. For instance in the case of *Mytilus californianus* studied in southern California, major growth rings were formed at intervals of 6–8 months, coinciding with spawning.

Mytilus edulis also shows 'microgrowth' bands in the shell (Richardson, 1989). These bands correspond to emersion times, as they do in other molluscs (see p. 87). Groups of mussels held in mesh cages at MTL show growth bands that are clear and well defined. In subtidal cages, however, growth bands are weak and show much variation between individuals.

Predators of mussels

A great variety of predators take advantage of the sessile food supply offered by mussels. In Yorkshire, England, *Mytilus edulis* at their upper intertidal limit can reach 20 years of age, but on the low shore they seldom reach more than 3 years because of attacks from predators. Dog whelks (*Nucella lapillus*) drill holes through the mussel shells. Starfish

(*Asterias rubens*) pull the shell valves apart and, by eating some mussels, loosen others. The shore crab *Carcinus maenas* and the edible crab *Cancer pagurus* crush the mussel shells. Various birds, such as purple sandpipers, turnstones, oystercatchers and others, also feed on mussels.

How important are predators in controlling mussel populations? Experiments at Lough Hyne, Ireland, suggest that mussels on the exposed outer coast are often free of predators, while those in sheltered waters may suffer extreme predation from crabs. When mussels were transferred to new positions within the shelter of Lough Hyne, very few of them survived for more than 2 weeks, while mussels moved on the exposed outer coast survived indefinitely. Within the lough, the crabs *Carcinus maenas* and *Liocarcinus puber* broke open and ate many mussels, mainly by night. A few starfish, *Marthasterias glacialis*, also attacked the mussels. Many prawns, mainly *Leander serratus*, visited the mussels as though scavenging, especially when the crabs were feeding. However, mussels can grow large enough to resist these predators. When various sizes of *Mytilus* were set up in wire cages, large mussels were immune to all sizes of *Carcinus*, while small mussels were vulnerable even to small crabs (Ebling *et al.*, 1964).

Immunity from predators is not, however, found on all exposed coasts. The controlling effect on the lower limit of mussels by predators has been well demonstrated on exposed shores in Washington State for *Mytilus californianus*, in New Zealand for *Perna canaliculus*, and in Chile for *Perumytilus purpuratus* (Paine *et al.*, 1985). In each case, experimental removal of predatory starfish allowed the mussels to extend their distribution downshore. We discuss these effects further on p. 173, but should note here that other predators may also be important (Okamura, 1986). For example, on the Pacific coast of North America, the crab *Pachygrapsus crassipes* takes individuals from the edge of patches, while visiting shoals of fish, the surfperch, may completely remove mussel colonies. These fish shoals are irregular visitors, sometimes visiting less than once a year, but they may play a part in determining long-term diversity. When examining the effects of predators, therefore, it may be necessary to carry out experiments over prolonged periods.

Barnacles

The acorn barnacles are like modified shrimps lying on their backs and enclosed within a series of calcareous plates into which they can withdraw for protection. They form a part of the crustacean subclass Cirripedia, so they have six pairs of two-branched ('biramous') legs. By kicking these they catch planktonic food. Their shell has a base which is membranous in the common intertidal species but calcareous in sublittoral and some low littoral species. All cirripedes have a characteristic planktonic larva called a nauplius.

There are four very common intertidal barnacles in north-west Europe. *Semibalanus balanoides* is found on most shores. *Elminius modestus* is an import from Australasia (p. 121), and most common in shelter. The two *Chthamalus* species, characteristic of exposed shores, have only recently been distinguished from each other: the original *Chthamalus stellatus* has now been divided into the low-shore species *C. stellatus*, and the high-shore species *C. montagui*.

Breeding and development

Unlike many intertidal invertebrates, barnacles do not liberate their gametes into the sea, but reproduce by internal fertilization. They have a very long extensible penis, which is inserted into the mantle cavity of neighbours. Cross-fertilization can only occur between barnacles within penis range of each other, and on any such occasion one of the pair acts as male and the other as female. What is the effective range? Only a reasonably dense population can propagate itself. For *Elminius modestus* the maximum distance apart for fertilization is 3–5 cm. The used penis is shed at the moult following an act of fertilization, and a new one is grown before the next occasion.

The times of breeding and rates of development vary greatly with latitude. In Tromso, Norway, most fertilization of *Semibalanus balanoides* occurs in early October; in Wales most occurs in early November; and in southern England fertilization is as late as early December. Dates vary with position on the shore, too—individuals growing at MHWN achieve 50% fertilization about 2 weeks before those growing at MLWN.

These variations are probably related to temperature differences. Each species has an 'optimum' for rate of development and hatching. *Semibalanus*, for example, which is a northern species whose southward limit reaches only just into Spain, breeds over winter and has an optimum of 14°C. *Chthamalus* is a more southern species which reaches down to west Africa and has an optimum of 31°C.

Settlement of barnacle larvae

The fertilized eggs hatch to produce the stage 1 nauplius larvae and further moults convert these to the final stage 6 nauplius larvae which are much larger. These moult in turn to give the cypris larva, which has a bivalve carapace and looks rather like a miniature ostracod. In some species the nauplii are brooded in the mantle cavity, but usually they escape into the sea and are free-living.

The settlement of the cypris larvae depends on whether they can find a suitable substrate, and can be postponed for nearly 2 weeks (Knight-Jones, 1953). At settlement, the cypris larva walks about on a surface which it is exploring and swims from time to time. If it finds a barnacle of its own species, or even just the base of one, it changes its behaviour: it walks more slowly with more frequent changes in direction. After about 30 min exploring the surface depressions it settles down in one of them facing as nearly as possible towards the incident light, and extrudes attachment cements which glue it to the substrate. Metamorphosis follows and it becomes a barnacle of the adult form 24 hours later in *Semibalanus balanoides*, slightly less in *Balanus crenatus* and within 4 hours in *Elminius modestus*.

When settling, cyprids strongly favour substrates that already have their own species attached. In laboratory experiments, *Semibalanus* larvae settled preferentially on the shell valves of live or dead *Semibalanus* adults. A few settled on shell valves of live *Balanus crenatus*. Fewer still settled on shells of *Elminius*, and very few settled on *Mytilus* shell valves which had no trace of barnacles. This settlement behaviour is known as aggregation, and presumably results in populations dense enough for fertilization.

What and where is the substance that promotes aggregation? Clean surfaces are unfavourable for settlement, but can be made attractive by soaking them in extracts of the tissue of barnacles or other arthopods. This substance resists severe chemical treatment and is probably a protein, now called 'arthropodin'. Arthropodin in solution does not induce settlement, but only when spread over a surface: slate panels treated with extracts of *Semibalanus* prove very favourable for settlement by the cypris larvae, whereas untreated panels in sea water prove unfavourable. Untreated panels placed in extract solution are at first unfavourable but slowly become weakly attractive, as if the arthropodin is slowly adsorbed by them. How does such a molecule stimulate the sensory receptors of the cypris larva? Although there are various theories, there is, as yet, no conclusive answer (Gabbott and Larman, 1987).

Growth and life span

Even within their protective calcareous plates, adult barnacles continue to moult. Well-fed adults of *Semibalanus* may achieve up to 15 moults/ year, increasing in size each time.

As they grow, adults adjust their orientation so that the cirri face towards the prevailing water flow. This gradual change of orientation by the adult barnacle is recorded by a spiral twist of the radial canals in the base of *Balanus improvisus*, well shown from underneath in barnacles attached to a glass plate.

In *Semibalanus*, growth is limited to spring and summer, but varies greatly with height on the shore (Fig. 6.5). We can follow two popula-tions on the Isle of Man to give an idea of life histories. Cypris larvae settled in the spring, and at the end of the year individuals at MTL spawned, releasing larvae in the spring exactly 1 year after settlement. Some of them then died but others grew slowly and spawned again the following spring, after which the entire population died. In contrast, individuals which settled at MHWN grew fast but did not spawn in their first year. They spawned well in the spring of their second year, and survived and spawned again each year even up to their sixth year after settlement. In their old age, however, some became sterile and growth was slow.

Fig. 6.5   Size distribution of *Semibalanus balanoides* on the Isle of Man. At MTL (open circles), most of the population remains small. The majority die in their third year. At MHWN (solid circles), mean size is greater, growth rate is higher, and many individuals live for 5 years or more. (After Moore, 1934.)

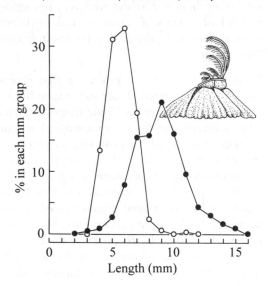

Barnacles tend, in settling, to maintain a distance of about 2 mm from previous settlers, allowing them to grow initially without touching the growing edge of another barnacle. However, after this initial period growth is greatly affected by competition between individuals (discussed on p. 167). Intraspecific competition changes the adult form dramatically, so that dense settlement results in tall, 'dog-tooth' shapes instead of the more normal 'volcano' shape (p. 163).

Distribution and climate

Population densities of barnacles vary dramatically in relation to exposure. Many species are more abundant, and extend further up and down the shore, at wave-exposed sites than in shelter. There may be many reasons for this. The much greater quantities of fucoids in shelter may dislodge recruits, bruise the adults or overgrow them, or may repel them in some other way. The swash of waves at a wave-exposed site no doubt protects barnacles from desiccation during low water. In addition, barnacles are eaten by the gastropod *Nucella lapillus*, which is discouraged by extreme wave action.

As we have mentioned above, different species also show varying distributions in relation to latitude. The fact that *Semibalanus* is more northern in distribution than *Chthamalus* suggests that temperature might control distribution in some way. However, experiments in which adult *Chthamalus* were transferred to cold sites on the east coast of England, where the species is not normally found, demonstrated that they could survive for two winters, and could spawn.

Over the period 1940–50, *Chthamalus* increased in abundance in Britain while *Semibalanus* declined. The air and sea-surface temperatures around the British Isles appeared to become warmer during this period. This might account for the observed change, but other explanations are quite possible. For instance, there was also an increase in the incidence of the parasitic isopod *Hemioniscus balani* in *Semibalanus*, as though it had in some way become more susceptible. It also seems unlikely that low temperature could be the factor that limits *Chthamalus* directly because it is found in the far north of Scotland, on Fair Isle and on the Shetland Islands. Possibly it requires Atlantic rather than coastal water, and it may be that its distribution is related to distribution of planktonic food suitable for the larva.

In the 40 years since 1951, sea temperature near Plymouth has declined again (Southward, 1991). From 1951 to 1975, *Chthamalus* declined and *Semibalanus* increased, but from 1975 to 1990, this trend has been reversed (Fig. 6.6). In summary, there is probably a relationship between abundance of the two species and sea temperature, but many other factors contribute to the changes. The long-term trends can be confused with short (10 year) cycles, and an overall influence of the 'greenhouse' effect may be involved.

The invasion of *Elminius modestus*

The Australasian barnacle *Elminius modestus* was first discovered in Europe in 1945, on plates exposed to fouling near Southampton, England. The adults probably came over on ships' bottoms in 1939 and larvae spread from there locally. In 1946 more *Elminius* were discovered at Plymouth, England, and then on the Dutch coast and in the Seine Estuary, France. *Elminius* competes for space with *Semibalanus*, and has taken much of the space vacated by this species in western Europe. It also thrives in dirty harbours and muddy estuaries, where other species are uncommon.

Fig. 6.6   Trends in sea temperature and relative abundance of barnacle species near Plymouth, England. The upper graph gives smoothed annual means of temperature, and the long-term trend for 1950–88 (straight line), which shows an overall decline. The four barnacle species are *Chthamalus montagui* (*Cm*), *C. stellatus* (*Cs*), *Semibalanus balanoides* (*Sb*) and *Elminius modestus* (*Em*). The lower graph shows the barnacle index (ratio of total *Chthamalus* spp. to total barnacles) at high shore (MHWN–MHWS) and low shore (near MLWN). Straight lines show regressions. The index on the high shore remained constant. On the low shore it fell until the early 1980s, but has since shown a rise. (After Southward, 1991.)

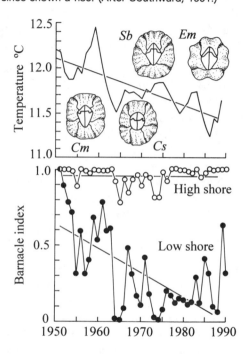

Bryozoans and other encrusting groups

On the upper parts of rocky shores, there is usually a very low diversity of sessile encrusting fauna. Here much of the fauna, apart from the barnacles, is mobile and able to retreat from the rigorous conditions of emersion into humid refuges. Low on the shore and in the sublittoral, however, sessile animals are common, particularly on macroalgae and the undersides of rocks. Many bryozoans, sponges, tunicates and hydroids are common in specific microhabitats on sublittoral boulders.

For many encrusting groups, ecological studies have been hampered by problems of identification and taxonomy. Sponges, in particular, are abundant and diverse in the sublittoral but are hard to identify. Recent work has concentrated upon species living not on the rocks but on macroalgae—the 'epiphytic' species—and of these the Bryozoa have attracted most attention (Seed, 1985).

Environmental effects on adult encrusting animals

Current speed and siltation have major effects on most marine organisms, including the encrusting fauna (p. 41), and many epiphytic species show fastest growth in clean, fast-flowing water. In Strangford Lough, Northern Ireland, *Fucus serratus* which experiences high current flows has abundant faunal epiphytes, particularly the two bryozoans *Flustrellidra hispida* and *Alcyonidium hirsutum*, and the hydroid *Dynamena pumila*. Experiments in which algae were transferred from their normal clean, fast-flowing water to water with a slow flow and a high silt load gave slightly different results for the bryozoans and the hydroid. After 2 months, the bryozoans had grown both at the original (fast-flow) site and at the new (slow-flow) site, but there was much *more* growth at the original site. Hydroid colonies increased in size at the fast-flow site but when moved into slow-flowing turbid water decreased to about 50% of their original size.

These effects may be 'direct' in the sense that currents alter food and oxygen supply and affect delicate feeding mechanisms, or they may involve 'indirect' effects which alter the balance of community structure by affecting, for instance, predators or competitors.

The distribution of epiphytic animals on their plant hosts

Encrusting organisms are seldom randomly distributed on their host plants. A classic example is that of laminarians, in which the holdfast supports an immensely diverse fauna, whereas the fronds often have a quite restricted number of epiphytes. In *Fucus serratus*, the fronds have recognizably convex and concave surfaces, and there is a general tendency for animals to be more common on the concave side. The hydroid *Dynamena pumila*, especially, shows a higher abundance on concave than convex surfaces, and the same is true for sponges and tunicates. Such distributions could be determined by settlement preferences of the larvae, since some have been shown to be able to respond to small differences in curvature. Equally, however, they could be the result of differential survival and growth in response to differences in microhabitat: eddy currents may aid feeding on the concave side, or abrasion may be less there in comparison with convex surfaces.

Distribution also varies with distance along the fronds of the algal host. Near the base of *Fucus serratus*, tunicates and sponges are abundant. On the central parts of the fronds occur the majority of bryozoans, hydroids and spirorbid worms. At the distal ends the bryozoan *Electra pilosa* dominates. Why are there these microzonation patterns? There is a series of physical gradients from the base to the tip of an algal thallus, since the base will be within the benthic boundary layer while the tip may lie in the mainstream flow, or will at least extend towards it (p. 108). The algal fronds will, of course, also have their own boundary layer. Besides these physical gradients, the algae produce chemical gradients which are of the utmost importance to epiphytes. Secretion of antimicrobial substances such as tannins varies along the length of the thallus. Maybe of more significance, the location of new algal tissue varies with position. In fucoids, new cells develop at apical meristems, so new tissue arises distally (p. 61). In laminarians, cells develop at a meristem between stipe and frond, so that distal tissue is actually the *oldest* (p. 65).

Some species of bryozoan select the youngest regions of their hosts to settle, and it may be that this allows them to reduce competition for space. *Electra pilosa*, for instance, settles preferentially near the growing tips of *Fucus serratus*, and *Membranipora membranacea* settles near the meristem of *Laminaria saccharina*, at the junction of stipe and blade (Fig. 6.7). Note that settlement near the tip of *Laminaria* would allow the colonies to be rapidly abraded away as the frond itself erodes.

Fig. 6.7   The distribution of two bryozoans on their host algae. Solid lines show adult colonies. Dashed lines show ancestrulae (the first zooids that develop from a settled larva). Algal meristems (growth regions) are arrowed. *Membranipora membranacea* settles near the ('intercalary') meristem of *Laminaria saccharina*, and as the alga grows, adult colonies move towards the tip of the frond. *Alcyonidium hirsutum* settles near the (apical) meristems of *Fucus serratus*, so that colonies are 'left behind' in the middle of the frond. (After Seed, 1985.)

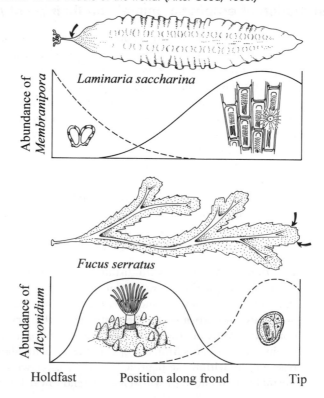

Spatial competition in bryozoans

When algae are colonized by large numbers of epiphytic animals, the colonies meet as they grow in size. Intense competition for space may then take place. Some sponges and tunicates can secrete acidic compounds that deter competitors and predators. These 'allelochemicals' can usually prevent settlement by other invertebrates: tunicate species that produce acid secretions are themselves free of epifauna or epiflora.

Competition between encrusting species is often manifested as 'overgrowth'. When they meet, several responses are possible (Stebbing, 1973). In most cases, when bryozoan colonies meet on *Fucus serratus*, growth ceases at the meeting point and continues in areas where there is no competition for space. There is never any overgrowth of one colony by another of the same species, and indeed, very infrequently two colonies of the same species may fuse together. The bryozoan *Electra pilosa* produces long calcareous spines at sites where the colony meets another species. In some cases these spines appear to prevent overgrowth, but two gelatinous species of bryozoan, *Alcyonidium polyoum* and *A. hirsutum*, usually grow over *Electra* in spite of this protection, covering the entrances to the individuals ('zooids') and presumably killing the edge of the colony. The two *Alcyonidium* spp. do not overgrow each other—when they meet, growth stops. *Alcyonidium* spp. and another gelatinous bryozoan *Flustrellidra hispida* also overgrow the hydroid *Dynamena pumila*, but this probably has little direct effect on it because its feeding and reproductive units ('thecae') are borne on erect stems which protrude above the overgrown ones.

Neighbouring colonies of bryozoans also interact at a distance, because they are feeding upon the same supply of planktonic food. In laboratory experiments, colonies of *Electra pilosa* reduce the growth of neighbouring colonies (Okamura, 1992). This effect is brought about by a reduction in feeding: upstream neighbours deplete the supply of food. In contrast, neighbours of different species such as *Alcyonidium hirsutum*, which are much larger than *Electra*, actually *increase* the amount of food captured under conditions of moderate water flow. This is presumably because, as shown by dye streams, the larger species alter the flow regime and direct more water towards the smaller tentacles of *Electra*. The enhancement of feeding may be termed facilitation. It is not as yet known whether such facilitation is common in suspension-feeding communities, or whether competition for available food is the overriding force.

Predation on bryozoans

The importance of predation in controlling abundance of epiphytic organisms is not well understood. Since most encrusting animals are colonial, asexual reproduction can often make good the damage done by browsing nudibranchs, fish and pycnogonids, as long as colonies are subject only to 'partial predation', i.e. some part of the colony survives. Nevertheless, predators can consume a large proportion of available standing crop, as shown in a study of the effects of a small nudibranch, *Doridella steinbergae*, on *Membranipora* spp. *Membranipora* colonies are most dense in central and distal regions of the kelp *Laminaria saccharina*, and these coincide with maximum population densities of the nudibranch. The *Membranipora* grows by budding off new zooids at the edge of each colony, at a rate of 22 000 zooids/plant/day. The mean density of *Doridella* is 238 individuals/plant. If the nudibranchs eat zooids at a rate that would keep the bryozoan at the same size, this would allow them *c*. 92 zooids each/day, more than enough to maintain them.

It is apparent that *Doridella* must make a very significant impact on the bryozoan population. Nevertheless, the bryozoan is not without its own defences. In response to the presence of *Doridella*, *Membranipora* can begin the rapid production of long spines at the periphery of the colony. These spines are thought to deter predators, just as those in *Electra* (p. 125) are thought to deter competitors.

It is important to note that the *Membranipora–Doridella* interaction was studied on kelp, which regenerates its blades frequently. Because of this regeneration, the area available for colonization changes rapidly depending upon the balance between growth at the meristem and necrosis at the tip of the frond. Growth rates of the bryozoans are high and predator life cycles are probably geared to coincide with those of their prey. On longer-living algae and on hard substrates, the situation may be quite different.

Polychaetes

Suspension-feeding polychaetes are more often associated with sedimentary than with rocky shores. But on some exposed rocky shores, where there is a good supply of sand, some sedentary polychaetes, such as *Sabellaria alveolata*, may be so abundant as to form a continuous coating on boulders and low-shore platforms. We discuss this species first, and then move on to *Spirorbis*, whose tiny coiled tubes may cover weeds and rocks.

Sabellaria and reef formation

Sabellaria alveolata constructs its tubes of sand or shell fragments, and may form reefs in the region below MLWN. In the Severn Estuary, England, there are extensive sublittoral populations even where the water is very muddy. Extensive colonies on clean, sandy beaches are more common, and form a honeycomb over the surface of boulders, varying greatly in size from year to year. The physical action of storms and varying coverage by sand often diminishes the area covered, but settlement of larvae and subsequent growth lead, in some years, to rapid restoration of the colonies.

In the bay of Mont St. Michel, France, colonies of *Sabellaria* form reefs over 1 m in thickness, but then suffer drastic erosion. New colonies develop on the eroded base of the reef, growing spectacularly by an average of 50 cm in 1 year. At this site the colonies go through cycles of growth and decay lasting about 10 years (Gruet, 1986).

Observations of the larvae of *Sabellaria alveolata* in the laboratory show that when able to metamorphose, after weeks of pelagic development, rough water with suspended sand grains encourages settlement. The major factor inducing metamorphosis, however, is contact with the tubes of worms that have already settled: the larvae need to contact the cement used in constructing them. Without this contact, larvae can postpone settlement for weeks. The behaviour of *S. alveolata* larvae thus parallels closely that of barnacle cyprids (p. 119).

The larvae of other reef-building species of *Sabellaria* are not so specific in their choice of substrates, and do not require the contact with cement needed by *S. alveolata*. It may be in their case that colony formation is determined more by the density of larval supply than by the behavioural responses of individuals to a marker produced by original colonies.

Spirorbids as epiphytes

Spirorbid worms live in small, coiled, calcareous tubes, closing the aperture with an operculum when they withdraw. There are numerous species in north-west Europe within the family Spirorbidae. One of the major aims that has dominated research into these polychaetes has been to account for the differing distributions of the species. There is some evidence that after settlement there is competition for food and space, but most attention has focused on the behaviour of the larvae. The spirorbids brood their larvae in the tube, or in a special chamber within the operculum. Adults of *Spirorbis spirorbis* kept in subdued light and then transferred to small bowls of fresh sea water in bright light liberate larvae in large numbers which can then be collected for study (Knight-Jones *et al.*, 1971).

Simple choice experiments showed that larvae of *Spirorbis spirorbis* from different sites in Wales have different preferences for host algae. Those from Pembrokeshire showed a significant preference for *Fucus vesiculosus*, the predominant algal species at the site of collection, to which the adults had been attached. Those from near Swansea, attached to the dominant seaweed *Fucus serratus*, chose *F. serratus* on which to settle. It seems that different selection pressures in the two habitats have altered the behaviour of the two stocks. To establish whether this was a temporary conditioning, or a more permanent effect, a third stock was examined, from Menai Bridge, where *F. serratus* and *F. vesiculosus* are both common, and both have abundant populations of *Spirorbis*. In this case, the larvae settled equally readily on both algae, indicating that the effect on selection of the host was not just a short-term conditioning, and suggesting that the stocks had diverged genetically.

If spirorbids can select host algae quite specifically, what are the chemical cues involved? There is no suggestion here, as in barnacles and *Sabellaria*, that larvae detect adult cement, though larvae of spirorbids are gregarious and tend to settle together. *Spirorbis spirorbis* larvae are attracted to extracts of their host alga, *Fucus serratus*, smeared on to settling plates (Fig. 6.8). The attractant is a product of the alga and not of bacterial metabolism because sterilized extracts are still attractive. Another species, *Spirorbis rupestris*, is also an epiphyte, but lives on the coralline alga *Lithothamnion*. Settlement of larvae is promoted by extracts of *Lithothamnion*, but when these extracts are boiled, the effect decreases, indicating that the attractant is heat labile. In contrast to these observations, larvae of an American species, *Janua brasiliensis*, can be induced to settle merely by the presence of bacterial films. The substances that promote settlement are thus probably diverse, and presumably related to the nature and habitat of the host algae.

Fig. 6.8   Settlement of larvae of *Spirorbis spirorbis* on panels treated with an extract of their normal host alga, *Fucus serratus*, and on control panels. (After Williams, 1964.)

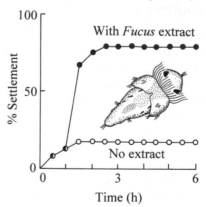

Sea anemones

By far the most obvious and common anemone in north-west Europe is the beadlet, *Actinia equina*, which ranges high up into the intertidal zone. This species has attracted much study in Britain, while in America and New Zealand, other common species have been investigated.

The food of sea anemones consists of a wide variety of crustaceans, molluscs, worms, other invertebrates and fishes. As predators, they utilize their nematocysts, but some also use ciliary currents to trap suspended particles and plankton. The differentiation between 'predators'—those animals that seek out and catch relatively large single prey—and 'suspension feeders'—those that trap many small animals, plants and inanimate particles in some rather more mechanical way— breaks down here.

Movement in sea anemones

Sea anemones are usually regarded as sessile, but in fact many of them can move around, using the musculature of the pedal disc. Ottaway (1978) followed the movements of *Actinia tenebrosa* at Kaikoura on the South Island of New Zealand. It proved impossible to tag anemones, so each individual was followed by establishing a fixed grid on the rock and mapping positions within this. Some individuals were apparently static for 2 years of observation, while others showed small moves of up to 50 mm. On some occasions, however, individuals showed sustained and directed locomotion. The greatest distance moved was 1555 mm, and the greatest rate was 210 mm/day. This sustained locomotion apparently occurs in response to some simple key stimulus such as physical injury, repeated desiccation, or wounding from intraspecific aggression.

Intraspecific aggression in the family Actiniidae

Actiniids are remarkable among sea anemones in that they possess specialized structures around the edge of the oral disc that can be used in aggressive interactions. These structures, the 'acrorhagi', are areas bearing powerful nematocysts. They can be inflated to form projecting organs. In *Actinia tenebrosa*, acrorhagi are expanded when individuals are brought together in the laboratory, close enough for their tentacles to touch. Usually only one of the anemones inflates its acrorhagi, becoming an 'aggressor', while the other does not, and can be referred to as its 'victim'. The aggressor then forces its inflated acrorhagi on to the column of the victim, and nematocysts are presumably fired, because when the aggressor withdraws, acrorhagial tissue remains attached to the column of the victim as if stuck there.

Field observations show that this aggressive response is triggered when a moving individual makes contact with a stationary one: the stationary one is the aggressor, and the moving one, or victim, moves away after being attacked. Not all intraspecific contacts lead to aggression, and this is now known to be determined at least partly by the degree of genetic relationship between individuals: aggression occurs principally between genetically dissimilar opponents.

On the Pacific coast of North America, *Anthopleura elegantissima* forms aggregations of individuals, each containing largely the offspring of asexual reproduction. These aggregations are regarded as clones, and each clone is separated from the next by a narrow space on the rock. Individuals at the edge of a clone fight with those of the neighbouring clone, thereby maintaining the 'no-man's land' between colonies, but those within a clone do not fight. Those at the edge of the aggregation have very large acrorhagi, but probably never develop gonads, and are termed 'warrior' individuals (Fig. 6.9).

Fig. 6.9 Acrorhagi of the sea anemone *Anthopleura elegantissima*, from the Pacific coast of North America. In the centre of clonal aggregations, acrorhagi (arrows) are small and there are few per individual. At the borders between clones, there are many, large acrorhagi. (After Francis, 1976.)

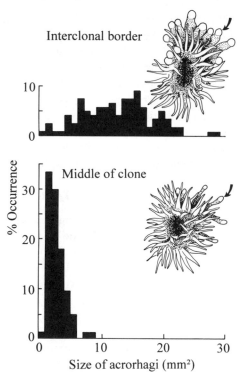

In the European species, *Actinia equina*, clones are not aggregated in this fashion. Aggressive interactions are similar to those shown by *A. tenebrosa*. In this case those individuals with red-brown columns are dominant over those with green columns, which seldom show any aggressive response at all. Among those with red-brown columns, aggressiveness varies (Brace and Reynolds, 1989). Enzyme electrophoresis has established that there are three genetic morphs: an upper-shore morph with red or pink pedal disc; a mid-shore morph, also with red or pink pedal disc; and a low-shore morph with a green pedal disc. Of these, the upper-shore and mid-shore morphs are most aggressive, and the lower-shore morphs least so. This dominance ranking may be reflected in distribution, but there have yet to be conclusive studies showing this.

Experiments to investigate suspension feeders

Many suspension feeders have the advantage (to investigators) of occurring in large numbers. Set against this, many are rather small and require microscopes for examination.

The distribution of Bryozoa on algae

The most common faunal epiphytes on macroalgae are usually Bryozoa. *Alcyonidium*, *Flustrellidra*, *Electra* and *Membranipora* are relatively easy to identify. Their distribution on particular host species forms a simple project. Distribution on each plant relates often to age of the plant: compare *Fucus serratus* (grows at the tips) with *Laminaria* (grows at the base of the frond).

The distribution of barnacles

Different species prefer different degrees of wave exposure and tidal level. This study could involve population structure (based on size), and can be related to laboratory experiments on tolerance of desiccation and of rapidity of feeding in response to wave splash or water cover.

Feeding in barnacles

The feeding responses of several species can be examined in the laboratory by observing their behaviour in still and in moving water. A water current can easily be produced by allowing water to run down a gutter from, say, an aspirator. To take this study further, the filtering mechanisms can be examined by killing a few individuals, removing the animals from the shell plates, and then examining the cirri under a compound microscope. Distances between spines give an indication of size of food caught.

Aggressive behaviour in sea anemones

Actinia equina is usually common. Individuals carefully detached from the rock will then reattach to small stones or shells. 'Fights' between individuals can then be examined either in the laboratory or in rock pools. Reactions between the various colour morphs, or between individuals from the same and different pools, can be examined. For a detailed laboratory study, nematocysts are easily seen in squashes of tentacles and acrorhagi.

Mussel distribution and behaviour in relation to habitat

Mytilus edulis is found on exposed shores and sheltered muddy inlets, allowing size composition of the populations to be compared. Are mussels found at intermediate exposures? To expand this project in the laboratory, feeding rates of different sizes can be compared by estimating the time taken to clear a standard concentration of graphite suspension from a fixed volume of water. This requires a spectrophotometer or colorimeter. Volumes cleared by different populations can then be estimated. The filtration rate changes with concentration of suspended sediment, so the project can become complex.

Inshore plankton

Sampling with a zooplankton net (100–200 μm mesh) and a phytoplankton net (30–60 μm mesh) can usually be carried out from any rocky promontory. Examination under a microscope gives an idea of food available to suspension feeders, and also allows the study of zooplanktonic suspension feeders (such as copepods) and larval forms.

7 Predators and their influences

The influence of predation on grazers and suspension feeders has been described at several points in the preceding two chapters, and it is evident that predation can be an important force structuring shore communities. But there are relatively few kinds of predators on rocky shores, reflecting their position at the top of the food web. Each is specialized for attacking a particular type of prey, so that some eat only the colonial sessile forms while others attack more mobile individuals. Some attack when the tide covers the shore while others wait until low tide. We begin by discussing the array of these various species (Fig. 7.1).

Fig. 7.1   The occurrence of some predators.

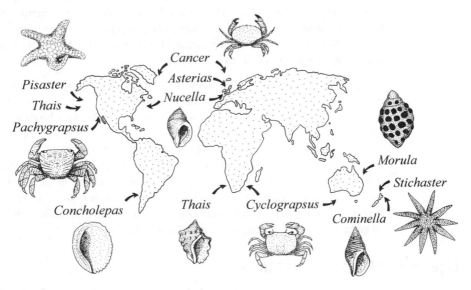

The distribution of predators

The most obvious predators on rocky shores, particularly on those exposed to wave action, are the gastropod molluscs known as whelks (Families Muricidae and Buccinidae). While these do not usually reach above MTL, the whelk genera are characteristic of every barnacle–mussel community. *Thais spp.* (= *Nucella* in Europe and parts of North America) are found world-wide, accompanied by others such as *Cominella* in New Zealand and *Morula* in Australia.

At lower levels on the shore, and sublittorally, starfish become abundant. *Asterias* and *Marthasterias* are common in north-west Europe, *Pisaster* on the Pacific coast of North America, *Stichaster* and *Coscinasterias* in Australasia. Most of these are predators of bivalves, particularly mussels.

The crabs are more hidden from view on many rocky shores, often because they migrate up and down with the tides, or lurk in crevices at low tide. The common families in north-west Europe are Portunidae (e.g. *Carcinus*) and Cancridae (e.g. *Cancer*). On the Atlantic coast of North America there are major fisheries for *Callinectes* (Portunidae) and *Cancer*. In Britain, there is an extensive fishery for *Cancer pagurus*, which landed 14 000 tonnes in 1984. On many other coastlines, crabs of the family Grapsidae are dominant: *Pachygrapsus* on the Pacific coast of North America, *Leptograpsus* in Australasia, *Cyclograpsus* there and in Africa. Grapsids tend to be active even at low tide, so crabs may be much more evident on these shores than in Europe.

At low tide level and sublittorally, the diversity of predators increases. Here there are many smaller forms, such as the nudibranch gastropods, which feed on sessile prey. Pycnogonids, polychaetes and nemertines may all be abundant. Below low water, octopuses may be quite common, and in Australasia, the Pacific coast of America and Africa they are probably significant predators of crabs.

The predators listed so far are always to be found on or near the shore, though many show vertical migrations. Two other important sets are the fish, many of which invade the shore only at high tide, and birds which appear only as the tide ebbs. Of the fish, gobies and blennies are universal, but other families are important on some shores. The weed-fish *Pseudolabrus*, for instance, preys on barnacles in New Zealand. Various bird species are active shore predators, and oystercatchers are voracious predators on mussel beds. Some species, such as herons, feed on the migratory fish fauna. Gulls, in particular, act as scavengers all over the world, eating dead or dying animals.

The effects of predators

There is no doubt that predators form a major part of littoral food webs. But do they affect the density and distribution of their prey? Or can prey populations reproduce rapidly enough to counteract the predator effect? Are there indirect effects? There is now widespread agreement that the feeding of at least *some* predators can have serious repercussions for prey species and for whole communities in the short term and in some areas. It is, however, much more difficult to generalize over wide areas.

Intertidal predators

There have now been many studies in which the effects of predators on intertidal communities have been examined experimentally, usually by excluding them. In Scotland and the Pacific coast of North America, removal of *Nucella* spp. allows increased survival of barnacles. In Chile, New Zealand and the Pacific coast of North America, removal of starfish allows increased survival and growth of mussels. On the Atlantic coast of North America, removal of *Nucella* allows growth of both mussels and barnacles. On the south-eastern coast of Australia, removal of the whelk *Morula* increases the density of most prey species, but to very different degrees in different habitats, showing that the effects of predation are patchy in space and time.

In each of these cases, primary carnivores are involved: their main prey is suspension feeders or grazers. The effects of predators have in fact been demonstrated in the marine intertidal zone on more occasions than in other ecosystems (Sih *et al.*, 1985). Their effects on density or survival of prey have often been shown to affect vertical distribution of prey (p. 25). For instance, on the Pacific coast of North America the major factor limiting downshore colonization by mussels is predation by starfish. Similarly, a major reason why the barnacle *Semibalanus balanoides* is often limited to the upper shore in north-west Europe is predation by the whelk *Nucella lapillus*.

Nevertheless, the effects of predators are much more complex than this brief summary might suggest. Predator species do not act in isolation, but within a community of primary producers, herbivores, detritivores, suspension feeders and other predator species. One example makes clear how complicated the interactions can be between the various species or guilds (a 'guild' being a collective term for species that exploit a resource in a similar way). In New England, the dog whelk, *Nucella lapillus*, is common on both exposed and sheltered coasts, but is only important in determining community structure in shelter. This is shown by experiments in which *Nucella* is excluded (Menge, 1976). On exposed coasts, exclusion has no effect on density of barnacles and mussels. In shelter, though, exclusion of *Nucella* allows barnacles and mussels to survive in much larger numbers than usual (Fig. 7.2). These take over the former 'free' space, which was utilized by algae such as *Fucus* species, microalgae and grazing limpets, and form a dense mussel bed. Overall it seems that the predatory action of *Nucella* is a major force normally maintaining a diverse community of algae and grazers.

This picture of *Nucella* as the sole predator that indirectly determines community structure on New England shores has been challenged recently on two counts. First, *Nucella* is far from being the *only* significant predator on sheltered shores. At high tide, the crabs *Carcinus* and *Cancer* are active, and their food includes the same prey as *Nucella*. Furthermore, large numbers of a fish, *Tautogolabrus adspersus* (the cunner) invade the shore at high tide and feed extensively on mussels and barnacles. Because these fish are rapid feeders, they may have a very significant effect on sessile prey even if they are much less numerous than the dog whelks.

Fig. 7.2   The effect of removing dog whelks, *Nucella lapillus*, from a sheltered shore at Grindstone Neck, Maine. Experiments were carried out in mid-shore areas normally dominated by *Fucus vesiculosus*, from which the algae were manually removed until March 1973. In the control area, some barnacles and mussels survived on the rock surface, but *Nucella* ate many of them. When *Nucella* was excluded, mussels alone came to dominate. When *Nucella* was excluded and mussels were then removed, *Fucus* sp. settled. (After Menge, 1976.)

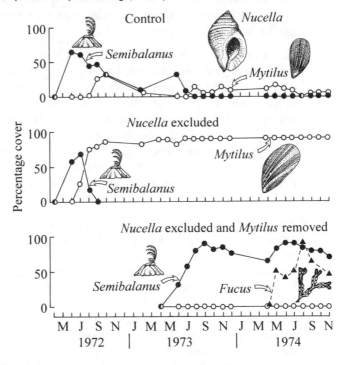

Secondly, when similar experiments are carried out at other nearby sites in New England, there is not always a detectable effect of *Nucella* on recruitment of *Mytilus* (Petraitis, 1990). Does this mean that *Mytilus* is not always the preferred prey of *Nucella*? The apparent preference for *Mytilus* in the earlier study may have been due to a bias brought about by the length of time it takes *Nucella* to consume *Mytilus*, relative to the time it needs to consume barnacles: prey that take a long time to consume are more likely to be seen and counted, and thus occurrences in the diet are artificially inflated. Recruitment of *Mytilus* in Petraitis' experiments was, in fact, closely related to another factor, the roughness of the rock surface—and increased roughness is well known to enhance settlement. The apparent *direct* effect of *Nucella* on *Mytilus* seen previously could therefore have been an *indirect* effect due to increased roughness caused by *Nucella* when it fed on barnacles, since the empty shells of barnacles create a very rough surface. Evidently, the relatively simple hypothesis first erected now needs considerable re-evaluation.

A later investigation has demonstrated that one of the complications is a great variation from place to place in the importance of *Nucella* as a predator (Menge *et al.*, 1994). We shall have more to say later about the geographical scale over which predators may be important (p. 185).

There is no doubt, however, that the overall influence of whelks and starfish *can* be dramatic. On the Pacific coast of North America, *Pisaster ochraceus* is responsible for removing the entire mussel population at particular heights on the shore, and for changing the community from essentially a monoculture to a diverse system of grazers and suspension feeders. Similar systems may operate in South America and New Zealand, although starfish are not so important on Australian shores.

The influence of the more mobile predators such as birds, crabs and fish has not until lately received as much attention as that of slow-moving whelks and starfish, probably because they are more difficult to manipulate. Many of these mobile predators act as secondary carnivores as well as eating grazers and suspension feeders: birds will take crabs and crabs will take whelks, so that the distinction between primary and secondary carnivores breaks down.

On the Pacific coast of North America, avian predators have variable effects on mussel recruitment. The use of cages that exclude birds but not invertebrate predators has shown that *Aphriza* (surfbirds), *Larus* spp. (gulls) and *Haematopus* (oystercatchers) reduce the recruitment of *Mytilus* at four sites out of six. Predation by birds is patchy and varies from site to site and over time. Social interactions are probably important here, since oystercatchers that feed on estuarine mussel beds change their feeding habits depending upon both the density of the food and the density of the feeding flocks (Goss-Custard *et al.*, 1981). However, because birds are rapid feeders, even sporadic visits can cause significant effects. The effects of fish upon Pacific-coast mussels are also sporadic and unpredictable. Mussel beds can be decimated by occasional visits from *Damalichthys* (surfperch), but if visits do not occur until mussels have reached a large size, the mussels are immune to attack. Some birds, in contrast, return regularly to particular areas to feed. In Lough Hyne, Ireland, predation by hooded crows results in a loss of nearly 25% of the mussel population each winter.

The effects of predation so far discussed have mostly concerned the density and distribution of the prey, although it has also been pointed out that important indirect effects on other species can occur. In addition, long-term effects of predation can produce evolutionary change in the appearance and life histories of prey species. Predators hunting by sight, such as fish, for example, may affect the distribution of different colour morphs of the winkle *Littorina mariae* (p. 93). Another example is provided by crab predation, which can alter the shell shape of dog whelks, *Nucella lapillus* (discussed further on p. 142).

Intense predation by crabs on two winkle species, *Littorina obtusata* and *L. mariae*, may have resulted in their use of different refuges (Williams, 1992). Juveniles of both species are attacked readily low on the shore (Fig. 7.3). *Littorina obtusata*, a long-lived species, avoids this predation by living on the mid shore, and has a spatial refuge because predation pressure from crabs there is low. *Littorina mariae* can exist on the low shore because it is an annual: it reproduces rapidly when crab predation is at a minimum, and has a temporal refuge when crabs have migrated offshore.

Fig. 7.3   Survival of *Littorina obtusata* when tethered on a sheltered shore at Sawdern, west Wales. On the mid shore, adults survived while juveniles were eaten by small *Carcinus maenas*. On the low shore, larger *Carcinus* were present, and both juvenile and adult winkles were soon eaten. (After Williams, 1992.)

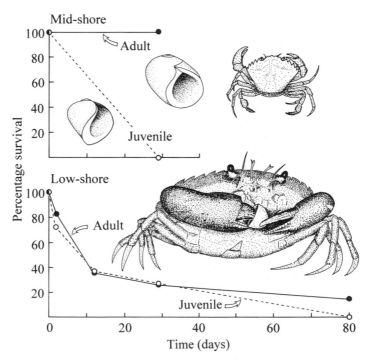

In summary, predation can have direct effects upon density and distribution of prey. It can also have indirect effects upon the balance of competition on the shore. Both effects can influence community composition. The scale of this influence is, however, variable, and the factors that determine whether or not a predator has a strong influence at any one site are not understood.

Subtidal predators

Because of the striking effect that grazing sea urchins can have upon subtidal flora, a great deal of interest has centred on the effects of the predators of sea urchins. Do such predators control urchin populations? If they do, does such predation release benthic macroalgae from grazing pressure and allow the development of kelp forests? Are sublittoral kelp forests thus dependent for their existence upon the presence of the predators of urchins?

On the Atlantic coast of North America, one of the major predators of sea urchins is the lobster *Homarus americanus*. Lobsters have been shown to be able to reduce urchin populations (p. 101), but it is evident that many other factors such as disease and predation by fish can also affect urchin densities. Whether kelp forests arise after the depletion of urchin populations is determined by a further set of factors, so that some authors at least have doubted whether the lobster should be regarded as a 'keystone' predator.

More studies have been carried out on the Pacific coast of North America, where a major predator of sea urchins is *Enhydra lutris*, the sea otter (Van Blaricom and Estes, 1988). Sea otters were once common from southern California round the North American coast to Alaska, and across to Japan, but hunting during the eighteenth and nineteenth centuries brought the species nearly to extinction. Only when protective legislation was passed did populations begin to rise again.

The otters capture mainly benthic invertebrates, especially crabs and sea urchins, then lie at the water surface and gain access to the flesh by pounding the prey on stones which they balance on their chests (Fig. 7.4). In some areas they also feed on fish, but this is not common in California. When sea otters move into an area, or are reintroduced, the density of urchins falls rapidly, and they retreat into crevices instead of being common on the open sea floor.

Fig. 7.4   A sea otter, *Enhydra lutris*, about to eat a sea urchin, *Strongylocentrotus* sp.

This brings us to the apparent simplicity of the subtidal communities in California: although there are many species present, the communities seem to be dominated *either* by the large kelps *Macrocystis* and *Nereocystis*, which form so-called 'kelp forests'; *or* they have little growth of macro-algae but are dominated by urchins—the so-called 'urchin barrens'. A continuing controversy surrounds the factors that cause the switch from kelp forest to urchin barren: is predation by sea otters, which reduces urchin grazing, the major cause of kelp regeneration? Or are there in fact many kinds of trigger that can cause this switch, the action of otters being only one of these?

It seems fair to say that there are two schools of thought here. One school quotes, first, the historical evidence that rising kelp dominance is associated with recent rises in otter populations; and, secondly, experimental evidence that when urchins are excluded, the kelp regenerates. This school labels the otter as a 'keystone' species, and says it is responsible for major changes in sublittoral ecosystems over wide areas. The second school points to evidence showing that switches between kelp forests and urchin barrens can occur where there are no sea otters; that storm-generated waves can cause great decreases in urchin density; and that storms can also destroy kelp forests by direct physical action. This second school therefore points to otters being merely one factor among a set of others.

The controversy can be simplified by saying that otters *can* affect urchin populations, and that in many cases reduction of urchins allows kelp growth. The argument is really about how general and widespread this effect is. Why do otters have a 'keystone' effect in some areas but not in others? This is presumably because sea urchins are affected by their local environment and may not be able to expand in numbers even when predation ceases. Similarly, kelps have their own habitat requirements, and can only grow if these are fulfilled, even when grazing pressure is lowered. It is the balance between these kinds of factors that needs to be understood.

Dog whelks

Dog whelks (*Nucella lapillus*) are slow-moving but tenacious, and provide the most accessible example of a predator on European shores. Many populations decreased drastically in the 1970s and 1980s, but are now recovering. *Nucella* reaches from Portugal to Novaya Zemlya, off the north coast of Russia. This distribution lies between the 20°C and the 0–1°C isotherms.

Feeding and diet

Nucella eats mainly barnacles or mussels, although it occasionally bores into shells of other molluscs, and even into its own species. When attacking mussels it bores through the bivalve shell, but when feeding on barnacles it is able to push the plates of the prey apart with its proboscis. Periods of boring with the radula alternate with periods of application of the 'accessory boring organ', a sucker-like projection on the foot. The secretion from this organ, in some unknown way, softens the shell, until a hole is bored right through the shell. After effecting an entrance, a narcotic is injected, followed by secretion of digestive enzymes. The dissolved tissues are then sucked up. Feeding is not a rapid process— consumption of one barnacle may take only a day, but to eat a mussel may take a week.

Food preference is greatly influenced by what has been eaten before: *Nucella* maintained on barnacles prefers barnacles, and *Nucella* maintained on mussels prefers mussels. Although this 'ingestive conditioning' takes a long time to develop, the frequency of occurrence of mussels and barnacles thus affects whether they are chosen as food.

Foraging behaviour on the shore varies widely. Feeding is constrained by the weather, so that heavy wave action on exposed shores, or desiccation in shelter, may prevent feeding for long periods (Burrows and Hughes, 1991). Nevertheless, under favourable conditions dog whelks avoid small prey and act as 'energy maximizers'. They can then ingest such large amounts of food that subsequent feeding is governed by the fullness of the gut: below a threshold level, they forage, but above it they remain in shelter.

Life history

In the spring, adults of *Nucella* tend to aggregate in large clusters, where they copulate. The females lay eggs in vase-like capsules, hundreds of eggs to each capsule. Only a few are fertile and will develop, the rest acting as food for the fertile eggs.

Eggs of *Nucella* are laid in April and May, and hatch as baby snails ('juveniles') without any planktonic stage in September or October. Egg capsules laid in low-level tide-pools may show high survival, but in situations which dry out during low tide many die. The first-year juveniles can be recognized as such until the following summer. Second-year 'immatures' can also be recognized, but third-years are indistinguishable from older shells and are classed together as adults. Growth of adults finally stops, the shells became thicker, and a row of teeth develops around the inside of the aperture lip.

Studies on young *Nucella* grown in a laboratory tank with an artificial tidal regime suggest that growth to a height of 10–15 mm occurs in the first year. These dog whelks were released at Plymouth, then were recaptured, measured and released again at intervals. Growth by a further 11 mm occurred during their second year and smaller growth as maturity was approached in their third year.

Variation in shell shape

Shells vary in shape from wide and thin-shelled on very wave-beaten coasts, to narrow and thick-shelled in shelter. Surveys of shell shape have been carried out at many sites, using the ratio of shell length/aperture length (L/Ap). In Wales, this ratio relates well to the rating of the shore on Ballantine's wave-exposure scale. On exposed headlands the shells are short and squat, with a ratio as low as 1.25:1. In sheltered bays, the shells are more elongated, with ratios as high as 1.5:1. This relationship holds in most cases, but in the Bristol Channel, England, dog whelks are unusually tall and narrow, with ratios as high as 1.7:1, and on the east coast of England dog whelks also deviate somewhat from the normal pattern.

Why do dog whelks show this extraordinary range of shell shape? Experiments at wave-exposed and sheltered sites suggest that we can explain two aspects of shape in terms of varying selection pressures. First, thick shells protect the snails from predation by crabs, as shown by experiments in which a crab (*Liocarcinus puber*) was put in a cage for 2 days with *Nucella* from both sheltered and wave-exposed sites: the *Nucella* from the wave-exposed site suffered overwhelmingly more damage (Ebling *et al.*, 1964). A similar picture was seen in experiments using the less-powerful crab *Carcinus*, so the thick shells that develop in shelter have probably evolved as protection against predation.

Secondly, the shells with large apertures have a large foot which allows the snails to cling better to the substrate, as shown by experiments in which the ability to hold on to a slate in a current of sea water was compared for individuals from exposed and sheltered sites (Kitching *et al.*, 1966). Individuals from the wave-exposed site held on much better than those from shelter. These experiments, however, do not explain why dog whelks in the Bristol Channel are so elongated.

There are also other fundamental questions that should be asked. In particular, what determines the variation in shell and body form of dog whelks? Are the characters under genetic control, or are they phenotypically determined by environmental factors? Conditions for genetic differentiation are probably most propitious for intertidal animals that lack any pelagic stage, such as *Nucella*. They tend to stay isolated in localised populations, and their genetic condition is preserved. *Nucella lapillus* is, in fact, differentiated into forms having different chromosome numbers—haploid numbers of either 13, or up to 18. In some places, such as at Roscoff, in Brittany, form 13 is found at very wave-exposed places and form 18 at sheltered places, but many intermediate forms also occur so that the importance of chromosome number is not clear.

Some features of dog whelks are under environmental control. In New England, hatchlings of *Nucella lapillus* newly emerged from eggs have the same size of foot whether they come from wave-exposed or from sheltered parents. Only when they grow into juveniles do they develop a foot of an area appropriate to their habitat. This is small for a sheltered site and large for a wave-exposed site, whatever their parentage. Transplant experiments in Devon, England (Fig. 7.5), show that when adults are taken from an exposed shore to a sheltered one, the progeny that survive all have the characteristics of sheltered shore populations, i.e. a small aperture and a thick shell (Gibbs, 1993). The conditions on sheltered and exposed shores can, therefore, determine the way that the shell develops.

Fig. 7.5　　Changes in the shape of *Nucella lapillus* in relation to environment. The original population (solid circles) from an exposed shore at Bude (north Cornwall) had large apertures and thin shells. When these snails were transferred to a sheltered shore in the Dart Estuary (south Devon), their progeny (open circles) developed thick shells and small apertures. (After Gibbs, 1993.)

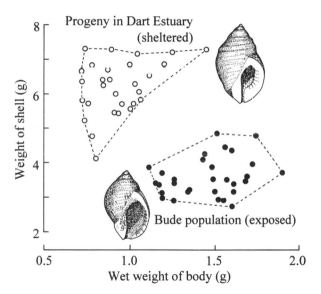

Another feature of the dog whelk shell besides general shape is the development of 'teeth' in the aperture. These teeth increase the thickening of the aperture lip and provide extra protection against predators such as crabs: a crab has greater difficulty in getting its chela into the aperture to pull out the body or in chipping away pieces from the shell aperture. The development of teeth occurs naturally as growth stops and the shell lip thickens at maturity. Sometimes a second row develops, as though growth had restarted and stopped again.

Both the growth of teeth and the growth of the shell as a whole can vary enormously in response to the presence of predators, as shown by a series of experiments carried out by Palmer (1990) in North Wales. He subjected dog whelks from wave-exposed and from sheltered sites to three different treatments: some were continually flushed with clean sea water, some with sea water flowing past *Cancer* sp. fed on fish, and some with sea water flowing past *Cancer* fed on dog whelks. In all these treatments, the dog whelks formed some apertural teeth over periods of 3 months, but the largest teeth were produced by dog whelks flushed with water in which crabs had been living and feeding. There was also a dramatic difference in growth rate between snails subjected to the three treatments: those flushed with water containing crabs feeding on dog whelks showed *no* growth, while those flushed with clean sea water or water containing crabs fed on fish showed increases both in shell length and shell weight. Evidently the dog whelks perceive the chemicals released by the crabs and their prey. Presumably they assess these in terms of 'risk' of feeding, so that in some situations feeding is suppressed.

Dog whelk evolution

Dog whelks are believed to have evolved in the Pacific, and to have spread into the North Atlantic in the Pliocene, during a warm period when it was possible to pass along the coast of the Arctic Ocean. They probably suffered severely, and were possibly completely eliminated from the North Sea area, by a series of ice ages. They might have been restocked from the north as the climate improved, even before the English Channel was open, but it is also possible that they may have originated from more than one source (Cambridge and Kitching, 1982).

The effects of tributyl-tin on dog whelk populations

In the 1970s and 1980s, dog whelk populations around south-west Britain showed a marked decline. This was accompanied by a rise in the incidence of what is called 'imposex': the addition of male characters to the female reproductive system. Females developed a penis and a vas deferens, and finally became sterile because the pallial oviduct was blocked. It took a series of studies over several years (Gibbs *et al.* 1988) to show that imposex was initiated by a chemical, tributyl-tin (TBT), used as a biocide in antifouling paints. This is so toxic that concentrations as low as 2 ng of tin/litre in early life can cause imposex, and the effects are not reversible. Over about 20 years, many local populations, especially those near harbours and marinas, became extinct. In 1987, legislation to ban the use of TBT on small boats was introduced, and since that time there has been some recovery.

Predators of dog whelks

Because of the strength of their shells, adult dog whelks are probably immune to attack by many predators, although a variety of birds, fish and crustaceans consume smaller individuals. Eider ducks swallow them whole and oystercatchers break them open, while crabs and lobsters crush the shells in their chelae. Juvenile dog whelks form a substantial part of the winter diet of purple sandpipers and other waders, while dog whelk losses in summer are mostly due to crabs. We discuss the foraging of crabs for dog whelks on p. 147.

Crabs

The most common intertidal crab in north-west Europe is the common shore crab, *Carcinus maenas*. It is found southwards as far as Morocco and northwards as far as The Faeroes and southern Iceland. It is also common on the Atlantic coast of North America from Nova Scotia to Virginia, and has been introduced to many other parts of the world.

The life history and seasonal migrations of *Carcinus*

Carcinus reaches sexual maturity when the carapace width reaches 15–30 mm in females and 25–30 mm in males. The female lays up to 185 000 eggs which are attached to long setae on her abdominal appendages or 'pleopods', where they develop for several months. They hatch to form planktonic 'zoea' larvae which moult five times, and then become 'megalopa' larvae. The megalopas sink to the bottom and metamorphose into juvenile crabs which are essentially miniature versions of the adults. They moult many more times, enlarging at each moult, and can become as large as 86 mm across the carapace.

Different sizes of adult *Carcinus* exhibit different distribution patterns and vary in their migratory behaviour. Smaller, younger crabs, especially those in their first year (< 34 mm carapace width) tend to live intertidally, hiding under rock or seaweed, or in crevices, at low tide. These young crabs are usually green in colour. Larger crabs move up and down the shore with the tide, and are found below low water level at low tide. The very largest crabs, which are often red in colour, live entirely below low water level. Superimposed on these changes with stage in the life history are the seasonal cycles: in winter, all individuals, even those in the smallest category, tend to move offshore so that the intertidal population may shrink to very low densities.

Activity rhythms of *Carcinus*

In the laboratory, without obvious environmental cues, the crabs show rhythmic activity, demonstrating that they have intrinsic internal rhythms. These rhythms can be analysed into two components, one of diurnal frequency (about 24 hours) and the other of tidal frequency (about 12.4 hours). The overall result in the laboratory is to produce peaks of activity during the high tides that occur at night. This rhythmic activity of *Carcinus* can be reset or 'entrained' by a change in conditions, such as a period of chilling followed by a return to normal temperature, or by other cues. These cues, known as 'zeitgebers', include hydrostatic pressure, low salinity and temperature.

Despite the neat patterns of activity that can be demonstrated in artificial conditions, trapping in the wild has shown that crabs are active enough, even at the time of low tide, to be caught in baited traps in similar numbers to those taken at high tide (McGaw and Naylor, 1992).

Predation by *Carcinus*

Carcinus eats a variety of intertidal invertebrates and even algae and detritus, but specializes in mussels (*Mytilus*) and dog whelks (*Nucella*). While mussels tend to occur in dense aggregations in which individuals are often of a uniform size, dog whelks occur in small numbers, often singly, and may vary in size even when clustered together. This means that the crabs have to use a variety of techniques to locate and open their prey.

One possible approach that could be taken by the crabs is to maximize the energy they obtain from the prey: if they were to follow this approach, they should attack the size and type of prey that yields the greatest amount of energy after allowance for energy spent in catching, manipulating, opening and eating it (Hughes and Elner, 1979). How does this theoretical picture fit that actually seen when the crabs feed on mussels and dog whelks?

We start by examining their approach to mussels. Crabs manipulate mussels in their chelae for 1–2 seconds, before accepting or rejecting them. The smallest mussels succumb to a single squeeze, but larger ones require a series of squeezes, and the largest can only be opened very slowly by breaking the shell along the edge, so that time taken to open the mussel increases with mussel size. In a series of aquarium experiments the crabs had many mussels of a range of sizes from which to choose (Fig. 7.6). For any one size of crab, the time spent in breaking the mussel shell and the time spent in eating increases steeply with increasing mussel size. As the mussel size increases, the prey value therefore increases, but it reaches a maximum (the optimum) and decreases again as the mussel shell becomes harder to break. When larger crabs are examined, time spent in breaking the shell of standard-sized mussels and time spent in eating decreases—so with increasing size of the crab the prey value peaks at a higher optimum level. Crabs do therefore select mussels with a high energy yield—assuming these are freely available.

When the crabs are placed with a limited number of mussels of various sizes, however, the situation changes. Optimal-sized mussels are still almost always accepted, but while suboptimal sizes are at first rejected they are accepted after several encounters. Here the decision takes account of availability. In addition to this complication, availability may be affected by competition: red crabs (from the sublittoral) have more powerful chelae than green crabs, prefer large mussels, and dominate green crabs in aggressive disputes.

Fig. 7.6   Consumption of mussels (*Mytilus edulis*) by shore crabs (*Carcinus maenas*) in laboratory experiments. When the crabs are given a free choice of size of prey, small crabs (open circles) choose small mussels, while large crabs (solid circles) choose large mussels. These curves coincide to some degree with the curves showing energy gained from various prey sizes. Energy yield is shown as energy in proportion to the time taken to break open the prey. For small crabs (open histograms), energy gained is high only for small mussels—they would probably *lose* energy trying to open large specimens. For large crabs (solid histograms), energy gained is higher for larger mussels. (After Elner and Hughes, 1978.)

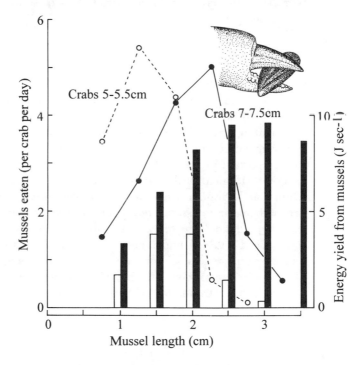

The requirements for handling dog whelks as prey are very different, because dog whelks are much less numerous than mussels, and the crabs cannot afford to reject all except the optimal. In laboratory experiments, *all* dog whelks encountered are attacked, but are discarded if not broken within 0.25–2.75 minutes. Small shells are crushed easily; larger shells are opened either by breaking off the shell apex or by inserting the master chela into the shell mouth to break the columella. This last method is strongly favoured against dog whelks from wave-exposed sites, which have wider shell mouths. Finally, the lip of the shell is chipped away, exposing some flesh, particularly on large dog whelks. This last procedure is probably the most economical for suboptimal individuals.

A somewhat similar situation is found when *Carcinus* feeds on winkles (*Littorina*). The smallest *Littorina saxatilis* and *L. nigrolineata* (0.5–1.0 cm high) are easily crushed. Medium-sized specimens (1.0–1.5 cm high) require some manipulation and often can not be broken. Attacks on large specimens (1.5–2.0 cm high) are only occasionally successful. When offered a mixed diet the crabs prefer the smallest sizes of both species, this being the most 'profitable' (in energy terms).

Carcinus can also extract *Littorina* without crushing the shell (Johannesson, 1986): using the smaller chela, it grips the operculum and foot of the snail, and pulls out the whole body. It is partly because of the use of this technique that winkles from sheltered shores, with small shell apertures, resist crab predation better than those from exposed cliffs, with wide apertures (p. 92).

Predators of crabs

Secondary carnivores in the littoral zone tend to be primarily the visiting vertebrates. Thus fish and birds, although only able to invade the shore at restricted times, are the major predators of crabs. At high tide, eels, bass, whiting and flounder move in with the rising water, and are joined by diving cormorants and shags. At low tide, herring gulls may be the most important predator. In Nahant, Massachusetts, gulls either stab the crabs with their beaks, and break them up, or consume them whole (Dumas and Witman, 1993). When *Carcinus maenas* and *Cancer irroratus* were tethered in mid-shore tide-pools, a number of crabs also suffered 'partial predation', i.e. they lost limbs but survived. Crabs survived better when tethered under algae, and also when they were allowed access to deep crevices; so, normally, they can find refuges from the gulls. The two species also show differences in their degree of cryptic colouring: *Carcinus* is better camouflaged than *Cancer*, and is not so vulnerable to gull attack.

Starfish

The radial symmetry of starfish, their five or more arms covered underneath with tube feet, and their ability to exert prolonged tension on their prey, fit these creatures as important predators, especially of bivalves and other shelled animals. They can evert their stomachs on to their prey, and some species are even able to insert the stomach into the gap between bivalve shells. They are not adapted for prolonged emersion, and occupy the low littoral and the sublittoral regions.

The most common species of starfish on the shores of north-west Europe is usually *Asterias rubens*. In the north are found *Crossaster papposus* and *Solaster endeca*, and in the south and west *Marthasterias glacialis*.

The food of starfish

Starfish prey on a wide variety of food, and this is well shown by a study using diving, carried out on the open sea coast of Sherkin Island, Ireland. Of 200 *Marthasterias glacialis* examined, 88 were feeding: 30 of them were feeding on detritus, 29 on barnacles and a few each on *Spirorbis* sp., various gastropods, *Lithothamnion*, the saddle oyster *Anomia ephippium* and *Echinus*. *Marthasterias* is thus quite omnivorous.

Fig. 7.7 A food web centred on the starfish *Asterias rubens*, on rocky shores in northern Europe. The width of the arrows represents the importance of trophic links. (After Menge, 1982.)

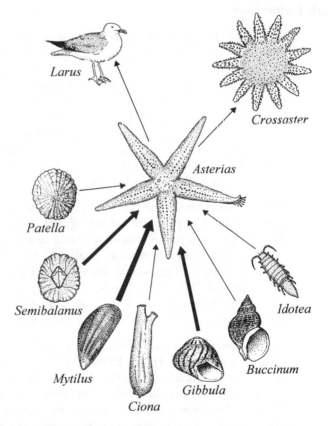

Asterias rubens is also very catholic in its tastes (Fig. 7.7), but in many situations specializes in attacking mussels (*Mytilus*). Although it normally lives near MLWS, it makes inshore migrations during the winter and depletes the lower mussel zone. On the Pacific coast of North America except in the most wave-exposed places, the bottom of the mussel zone is cut off sharply by the starfish *Pisaster ochraceus* (p. 26). *Pisaster* exhibits a size-selective feeding strategy which is very similar to that of the crab *Carcinus maenas* (p. 146): when presented with varied sizes of mussels, it chooses medium-sized individuals, probably maximizing its energy input.

On the Atlantic coast of North America species of *Asterias* invade the low littoral region from the sublittoral in summer and early autumn. Small individuals of *Asterias forbesi* prey on small *Mytilus*, while both medium and large individuals prey on medium-sized *Mytilus*. Large *Mytilus* are avoided. Calculations of energy expenditure and energy gain from the prey show that this behaviour only partially maximizes their energy intake: larger individuals apparently choose prey that is suboptimal in energy terms, but the reason for this is not clear.

Starfish behaviour

In Norfolk, England, *Asterias rubens* living in tide-pools comes up to the top of rocks at dusk, and goes down underneath them at dawn. What determines this behaviour? Observations in the laboratory show that the starfish move around during the night but stay still by daylight; during darkness they are usually distributed equally on the tops and the bottoms of the rocks. If the dark periods are reduced to 4 hours each, they come up and move around for 4 hour periods only, suggesting that light controls their activity directly. Illumination from below causes them all to move on to the tops of the rocks, showing that their response is to light and not to gravity.

Such experiments in the laboratory must, however, be treated with caution. The strong negative photoresponse of freshly collected *Asterias rubens* changes to a positive response after acclimatization to light. Added to this, responses are not the same in all field areas. In the western Baltic Sea at a depth of 15 m, *Asterias rubens* shows its greatest feeding activity during the day. This diurnal activity may be related to the reduction of the light intensity with depth, but may alternatively mean that separate populations do not all show the same behaviour patterns.

Similar problems occur with *Marthasterias glacialis*. In the shallow sublittoral of Lough Hyne, Ireland, numbers visible within a fixed area increase greatly by night. In the Mediterranean, activity of *Marthasterias* also occurs mostly at night (between 18.00 h and 09.00 h), but is not totally governed by light intensities: periods of inactivity of 1–3 days alternate with activity when deep-water individuals make feeding raids into shallow water (Savy, 1987). We do not at present understand what triggers these raids.

Some starfish are capable of limited learning, as shown by laboratory experiments with *Pisaster giganteus*, a shallow-water Californian species. When individuals were kept in the dark in separate tanks, each animal normally stayed on the wall. None of them responded to stimuli such as exposure to light. Eight trials were carried out in which the light was turned on for 15 minutes and at the beginning of this period a small mussel was placed on the bottom of the tank. The starfish came partly down on to the bottom of the tank, picked up the mussel and took it away. In the next set of trials the light was turned on and a mussel was given *after* the starfish had come down for it. This response took place rather more slowly than when the mussel had already been put in position. After nine such tests, all successful, further tests were carried out in which the light was shone for 15 minutes and no mussel was given. The starfish initially responded, showing that they had learnt to associate the light with a food supply. They began failing to respond within four exposures, and all failed by the ninth, showing that the learned response was retained only for a short time if not reinforced.

Nudibranchs

Some of the most colourful opisthobranch gastropods are the nudibranchs or sea slugs, which have no shell. There are 90 or so species of sea slugs in north-west Europe, some of them spectacular, but in fact the most common are not particularly striking. Most of the Doridacea are camouflaged or 'cryptic': *Archidoris pseudoargus*, for example, has a blotchy pattern in yellow and brown, while *Acanthodoris pilosa* is pale grey to purplish brown. In the Aeolidacea, *Aeolidia papillosa* is white to grey. These dull colours, plus their relative rarity compared with dog whelks and starfish, make them much less obvious than other predators. They cannot withstand desiccation, and occur under boulders or overhangs, where they attack sessile prey. Their relatives the tectibranchs, which have an internal shell, are found in similar places.

Food and feeding

Many nudibranchs are remarkably specific in their prey. *Archidoris pseudoargus*, for instance, lives almost entirely by eating one species of sponge, *Halichondria panicea*. *Aeolidia papillosa* normally eats the sea anemone *Actinia equina*, but it will eat other species in the laboratory, and its diet varies in different areas: on American coasts its major prey is the anemone *Metridium senile*. *Acanthodoris pilosa* is more catholic and eats a variety of encrusting bryozoans such as *Alcyonidium* and *Flustrellidra*. Some other sea slugs have wider tastes still, but almost all feed on bryozoans, hydroids, anemones, sponges or tunicates.

The feeding mechanism is, in most cases, rather similar to the grazing shown by herbivorous gastropods: a broad radula is applied to the sessile prey, then forced forward and up, scraping tissue into the mouth. *Aeolidia*, however, shows an extreme adaptation. The radula is narrow and the teeth have been reduced to only one per row, although each tooth does have a large number of cusps. There is also a pair of jaws.

When feeding on its anemone prey, it takes large bites from the column, and often kills more prey than it can ingest.

Defence

Unlike the majority of gastropods, nudibranchs have no protective shell into which they can withdraw when danger threatens. Nevertheless, they are far from defenceless, and indeed have adopted a much more active approach to defence than most other molluscs. There appear to be four major mechanisms by which they deter predators.

First, some species secrete sulphuric acid from dorsal epithelial glands when disturbed (Fig. 7.8). For example, *Onchidoris bilamellata* secretes a sulphuric acid solution of pH 2.0. The phenomenon is actually best demonstrated in some of the tectibranchs such as *Philine*, *Pleurobranchus* and *Berthella*, where stimulation of the dorsal mantle produces a secretion of pH 1.0.

Fig. 7.8   The mantle epithelium of the opisthobranch *Pleurobranchus peroni*. Acid-secreting cells produce sulphuric acid with a pH of 1.0, which deters would-be predators. (After Thompson, 1976.)

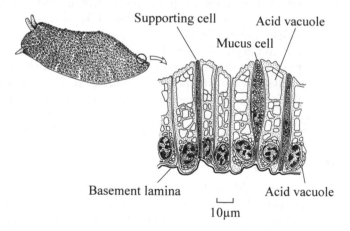

Supporting cell   Acid vacuole

Mucus cell

Basement lamina   Acid vacuole

10μm

Secondly, some nudibranchs have the amazing ability to extract nematocysts from their coelenterate prey, and then to use them in their own defence. This is common in the aeolids, which store the nematocysts in dorsal processes called 'cerata', which are often highly coloured.

Thirdly, dorid nudibranchs in particular have large numbers of calcareous spicules in their skin, often gathered into papillae which appear somewhat similar to the cerata of aeolids.

Lastly, but most widespread, nudibranchs have skin glands that produce a wide range of deterrent chemicals. These 'repugnatorial glands' are diverse in structure and little is known of the substances produced.

Overall, the various mechanisms appear to be very successful, because when tested with various fish predators, almost all nudibranchs appear to be distasteful. There are cases of fish consuming them, and some bird species, such as eider ducks, are known to eat them, but in general they are avoided. There has been much discussion about whether the bright colours of some nudibranchs advertise their unpalatability, but since even the cryptic species are unpalatable, no firm conclusions have been reached.

Reproduction and life history

Unlike prosobranchs, opisthobranchs are hermaphrodites, and they show reciprocal copulation. Nudibranchs usually mate in the late winter and spring, and at this time the characteristic ribbons of spawn begin to appear on the shore. Most species are thought to be annuals, and certainly the summer is a time when very few can be found on the shore. Because of their annual cycle, in which adults reproduce, giving rise to a planktonic larval stage, and then die, nudibranchs are particularly vulnerable to conditions in the open water as well as to those on the shore.

The common dorid, *Archidoris pseudoargus*, provides an example of how these processes are timed over the year. On the Isle of Man, the eggs laid in the spring take about a month to hatch. The veliger larvae are then liberated into the plankton, and young sea slugs are not found on the shore until the autumn. The juveniles mature over the winter, until they can mate and spawn in the spring, after which they die. Elsewhere, however, this species may be biennial (Todd, 1981).

Vertebrates

Most marine biologists visit the shore at low tide, and this is the time at which predatory birds gather there. Interactions of the fauna with predatory fish, however, take place both at low tide (in tide-pools) and when the tide is high. We consider the actions of fish first.

Littoral fish as predators

Littoral fish fall into several categories. There are the true residents, which are found only in the littoral zone, and remain there; the partial residents, which extend below low water in their distribution; the tidal visitors, which come in to feed as the tide rises; and the seasonal visitors, which come inshore to breed.

All of these categories may be important predators, but we know most about the residents. At Roscoff, in Brittany, the most common true resident is *Lipophrys (= Blennius) pholis*, the shanny, which is found in tide-pools and under weed over the whole shore (Gibson, 1972). Low on the shore and sublittorally, *Gobius paganellus*, the rock goby, is abundant, accompanied by *Crenilabrus melops*, the corkwing wrasse, as well as many others. Most rocky shores also have tidal visitors such as *Morone labrax*, the bass.

Even a casual search on shores in north-west Europe is likely to reveal the shanny, *Lipophrys pholis*. This fish has a wide-ranging diet, taking in quantities of algae and amphipods, but subsisting mainly on barnacles and winkles. Young *Lipophrys* eat many barnacle cirri—an example of partial predation, since the barnacles can presumably regenerate their appendages—while older fish tend to consume the entire barnacle, shell and all. *Crenilabrus* has a rather similar diet, but concentrates upon gastropods rather than barnacles. *Gobius paganellus* has a very wide-ranging diet indeed, eating few gastropods and no barnacles, but taking amphipods, copepods and ostracods. Each of these three predator species therefore exerts a different effect upon the shore community.

Lipophrys, like most other shore forms, breeds in spring and summer. The female attaches her eggs to the underside of rocky crevices where they are guarded by the male fish. He fans them to keep them oxygenated, and they develop after about 6 weeks into larvae that float in the plankton. Young fish return to the shore in the autumn at lengths of 4–6 cm. They may grow to a length of 16 cm, and live for as long as 16 years.

One of the ways in which *Lipophrys* is superbly adapted to life on the shore is in its activity rhythms (Northcott *et al.*, 1990). Experimental observation in the laboratory shows that even in constant conditions, *Lipophrys* exhibits peaks of activity at times of expected high tide. These peaks advance each day by about 12.9 h, slightly in excess of the normal tidal advance of 12.4 h. This endogenous rhythm gears the fish to be ready for feeding when the high tide covers the shore. The fish then leave their low-tide refuges and forage widely.

Birds as predators

There is a great variety of birds that feed on rocky shores. The only regular passerine is *Anthus spinoletta*, the rock pipit, but waders such as *Calidris maritima* (the purple sandpiper), *Arenaria interpres* (the turnstone) and *Haematopus ostralegus* (the oystercatcher) are common. *Ardea cinerea*, the grey heron, feeds on fish primarily at low tide, while *Somateria mollissima* (the eider duck) feeds on mussels by diving at high tide.

The role of these birds on rocky shores has been little studied, compared with the effort spent on observing the waders of muddy shores and estuaries. They have a variety of migratory patterns, but in general the waders move north towards the Arctic to breed in summer, and only arrive back on European shores in the autumn. Many stay for the winter, so that feeding pressure is presumably greatest then. This contrasts with the picture for fish, which probably feed less in the winter because they are poikilotherms.

The rock pipit is not a migrant, but even so its predation pressure on the shore diminishes in summer, when the surrounding terrestrial vegetation provides a rich fauna of insect prey. In winter, however, it spends more than 8 h each day foraging on the beach (Fig. 7.9). In south Cornwall, England, rock pipits forage particularly for the winkle *Melarhaphe neritoides*, and can eat an average of 33/minute. On some days, they can consume the amazing total of over 14 000 each. They eat chironomid larvae in large numbers, kelp flies, the isopod *Idotea* and the amphipod *Orchestia gammarellus* which is so common beneath decaying kelp.

Fig. 7.9   Size frequencies of the winkle *Melarhaphe neritoides* eaten by rock pipits (*Anthus spinoletta*) and turnstones (*Arenaria interpres*) at Porthcew, Cornwall, compared with a random collection. (From data of Gibb, 1956.)

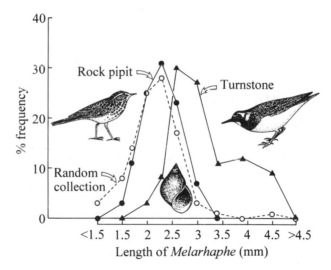

The purple sandpiper takes a wide spectrum of intertidal gastropods, from winkles to dog whelks, while oystercatchers take only the larger molluscs such as *Mytilus*, *Nucella*, *Patella* and *Littorina littorea*. Oystercatchers prefer *Patella aspera* to *P. vulgata*, probably because the shells of *P. aspera* are softer and easier to break and dislodge. In estuaries, they are important predators of mussel beds.

Of the larger birds, gulls and jackdaws are common sights on rocky shores, and some crows have evolved a method of feeding on whelks and mussels even though they do not crack them open with their beaks. In British Columbia, Canada, *Corvus caurinus* (the northwestern crow) feeds predominantly in the rocky intertidal (Zach, 1978). The crows pick up single specimens of the whelk *Thais lamellosa* from low on the shore, then fly towards the cliffs at the top of the beach. Here they fly up to a height of about 5.5 m, then drop the shell on to the rocks below. If the shell breaks, they remove the flesh and eat it, but if the shell does not break— as in the majority of cases—they try dropping it again. Each crow takes only large specimens of the whelk, and uses specific dropping sites, where a litter of broken shells accumulates. This sequence of behaviour fits neatly into the theory of optimal foraging: the crows collect only the largest whelks because smaller ones do not give them a large enough energy return to balance the costs expended in hunting and flying. Similarly, they fly up to about 5.5 m because this gives a good chance of breaking the whelk shell without expending too much energy.

In Lough Hyne, Ireland, *Corvus corone cornix* (the hooded crow) feeds on the mussel *Mytilus edulis*, the limpet *Patella vulgata* and the winkle *Littorina littorea*. The crows take only the larger specimens, and these mainly from the upper half of the shore, then fly to selected 'dropping sites' to smash the shells. Alternatively, some prey items are cached in vegetation near the shore before being dropped. Predation can be estimated by counting prey items at the dropping sites, for which the crows use both the rocky intertidal and local roads. It peaks in winter but falls to low levels in summer.

Experiments to investigate predators

Predation on gastropods by *Carcinus*

When *Carcinus* has been starved for 1–2 days, it will readily attack *Littorina* species in aquaria. Place individuals in separate aquaria and test reactions to species and sizes of prey. As well as making direct observations, leave overnight in the dark and record feeding preferences next day.

Distribution of *Nucella* in relation to prey

Dog whelks are limited to the lower shore, and predation intensity there tends to be high. Begin by counting the dog whelks feeding on each type of prey: various barnacle species, mussels and others. Are the dog whelks selecting prey according to their available proportions? Does this vary with size of dog whelk?

Variations in size and shape of *Nucella*

Using a variety of different shores—boulder beaches, exposed cliffs, sheltered shores—measure the shell height and the height of the aperture with callipers. Express aperture height as a proportion of total height. Does this vary from shore to shore? These observations can be made on the beach and dog whelks replaced in position.

Defence in nudibranchs

If a variety of nudibranchs is available, collect one of each species. Using a glass rod, stroke the dorsal surface of each slug, then apply wide-range litmus paper to the skin. The acid-secretors are easily detected. If aeolids are present, the nematocysts they have obtained from coelenterate prey can be seen by snipping off one of the cerata and viewing under a compound microscope.

Prey selection by *Asterias*

Asterias usually remains active in aquaria if kept for short times (1 day) before being returned to the shore. Its prey preference can be tested by presenting a variety of bivalves of different size.

Feeding in the rock pipit

Rock pipits usually feed more on the shore in winter than in summer. At Easter it is possible, using binoculars or a telescope, to observe individuals feeding and to count the feeding rate. This exercise requires a secluded site where human interference is low.

8 The functioning of rocky-shore communities

In Chapters 4–7 we have considered organisms on rocky shores in their separate trophic levels: first the algae, the primary producers; then the grazers and suspension feeders, the herbivores and detritivores; and lastly the predators at the top of littoral food webs. At each of these stages we have commented on trophic interchange between the various groups, but have not dealt with trophic organization overall. One way of approaching rocky-shore ecosystems is to examine this trophic organization, since it gives a view of the 'position' of the various species within the community. It is, of course, far from being the only way of viewing the community, but we start with this approach and then consider the various questions that such an approach raises.

Littoral ecosystems: food webs on the shore

The investigation of a food web can be carried out at many different levels. At an elementary level, it is possible to deduce the dominant species and their probable trophic positions from a combination of observation and reference to textbooks. On a more advanced level, the process requires measurement of biomass and rates of production, estimates of proportions consumed by the various components of the higher trophic levels, and measurement of the loss of energy from the system. It will be seen later that community structure can be affected both by the effect of higher trophic levels on those below ('top-down' effects) and by effects of lower trophic levels on those above ('bottom-up' effects). We begin with an elementary qualitative example and work up to more quantitative webs.

Two simplified food webs for the mid shore of south-west England have been derived from brief observations on student field courses. The web for an exposed shore (Fig. 8.1a) suggests that macroalgal production is mostly exported as detritus, whereas microalgal production is largely consumed by grazing herbivores such as limpets and winkles—a conclusion inferred from their abundance. The large populations of suspension feeders such as barnacles are supported by phytoplankton. Both herbivores and suspension feeders are attacked by predators such as crabs and dog whelks, and these predators are in turn consumed by 'top predators'

Fig. 8.1   Suggested food webs for (a) an exposed shore (Porthleven) and (b) a sheltered shore (Flushing), in Cornwall, England. The webs were derived and simplified from observations of abundance and diet of animals on the shore during field courses. Thickness of arrows suggests relative importance of trophic links.

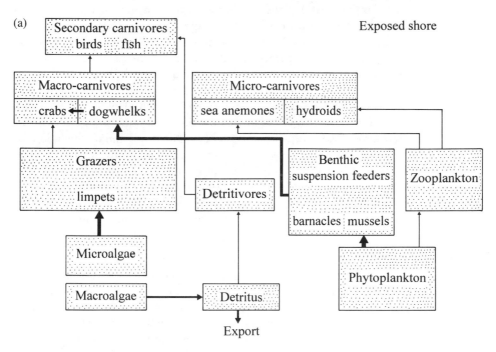

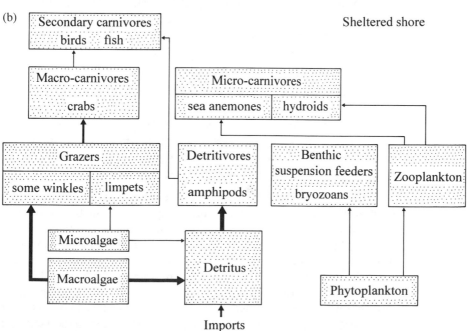

such as birds. This trophic structure leads to a classic 'pyramid of energy', in which production at each successive level declines drastically.

The web for the sheltered shore (Fig. 8.1b) differs particularly in that much of the algal detritus is no longer exported, but fuels a detritivore pathway. The carnivores therefore have a greater diversity of prey on which to feed because these include detritivores as well as herbivores and suspension feeders. Grazers that directly consume macroalgae are also present, so that much more of the macroalgal production is utilized on this shore, and grazing on microalgae is less significant. Although these two webs are crude and unsubstantiated in detail, they provide one way of looking at the two shores to see how their communities 'work'. One of their major drawbacks is that it is very difficult to estimate the influence of organisms not present at low tide, or present only seasonally. For instance, many fish migrate into the intertidal zone at high tide, while birds are often major predators only in winter. These kinds of hypothetical food web also need testing experimentally before any firm conclusions can be drawn, but they are useful at least in the generation of trophic hypotheses.

By restricting considerations to the mid shore, it is possible to produce a more quantitative analysis of energy flow (Hawkins *et al.*, 1992). For the Isle of Man, models of exposed shores show them to be net consumers (and therefore importers), while sheltered shores are net producers (and therefore exporters). However, such overall budgets do not forecast the fate of exported material, and in fact much of the detritus produced on the mid shore of sheltered shores merely migrates to the strandline or is metabolized by local subtidal organisms. In overall terms, sheltered shores collect detritus while exposed ones export it, because of the wave and current regimes.

In the low-shore/sublittoral zone of eastern Canada, measurements of biomass and production allow more detailed quantification (Miller *et al.*, 1971). Here the majority of primary production is channelled into detritus, and is then exported. A substantial proportion, however, is eaten directly by sea urchins. Detritus and phytoplankton supply food for brittlestars and mussels. Starfish are the major predators, while lobsters, crabs and fish are 'top predators'.

The approach to community structure via food webs generates a basic framework against which more difficult questions can be formulated. For example, we need to know not only how 'important' any one trophic link may be in determining the structure of the community, but in which direction it operates: do sea urchins control the seaweeds ('top-down' control), or does seaweed production determine sea urchin biomass ('bottom-up' control)? Do particular species exert more control than others (the 'keystone species' argument)? We need to know whether external conditions in the surrounding seas, where phytoplankton and pelagic larvae live, are important for shore communities. In particular, how does recruitment of larvae ('supply-side ecology') affect community composition? Before we tackle these questions, however, we must discuss whether the kinds of community that are found on rocky shores are persistent and stable, or whether they are subject to periodic changes. Only when we know how stable a community is can we hope to understand the factors that control its composition.

Stability on the shore

Long-term monitoring studies show, in general, that on sheltered shores the dense algal-dominated communities are relatively stable. On moderately exposed shores, where macroalgal cover is less, there may be considerable fluctuations in community structure from year to year. Exposed shores, in contrast, show little variation in overall community structure, but their small-scale mosaic patterns change frequently. We take these three sections of the continuous exposure spectrum in turn.

Sheltered shores

In north-west Europe, sheltered shores are usually dominated by fucoid seaweeds, of which *Ascophyllum nodosum* provides a blanket cover on the mid shore. This is a long-lived species, and the appearance of shores where it dominates may not change for decades. This stability is not universal, however, and we begin by describing a shore in the Severn Estuary, Britain, where considerable changes have taken place over 17 years (Fig. 8.2). The site is a steep limestone promontory, assessed as

Fig. 8.2   Abundance of algae and limpets over 17 years on a steep rocky shore at Portishead, Severn Estuary, England. Records from 1981–92 were taken from six fixed quadrats (each 0.25m²) at MTL (±1 m). Records from 1975–78 were taken on an adjacent slope and are estimates based on abundance scales. (From unpublished observations by C. Little.)

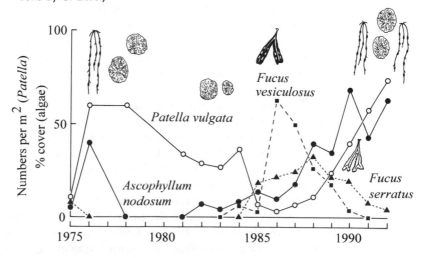

approximately exposure 6 on the scale of Ballantine. In 1976 the mid shore was covered by *Ascophyllum*, but in 1977–78 this virtually disappeared, leaving only fringes at MHWN and MLWN. A dense population of limpets, *Patella vulgata*, continued to inhabit the rock slope, until in 1984–85 their numbers declined. In 1986, dense settlement of *Fucus vesiculosus* occurred, while *Fucus serratus* grew gradually from 1985 to 1988. By 1991, both *Fucus* species had declined, while *Ascophyllum* showed some recovery, and the limpet population boomed. The site has therefore, in 17 years, seen a change from *Ascophyllum* cover to a limpet community on bare rock, then to dense *Fucus* cover, and back to *Ascophyllum* with limpets. The triggers promoting these changes are not well established. The original loss of *Ascophyllum* was probably due to storm action, since the plants were by then old (up to 18 years) and with long fronds. The decline of the limpet population was probably caused by winter temperatures as low as −13°C, which did not kill the animals, but made them lose their attachment to the rock. On a shore nearby but even more protected from wave action, changes over the same period have been very small. The change from *Fucus* species back to limpets does not seem to have any direct physical trigger, but it may be that since *Fucus* species live for only 4–5 years, their disappearance is due to the prevention of new spore settlement by the grazing of limpets. We can conclude even from this crude study that in such a situation both physical and biological factors can interact to cause long-term fluctuations.

Moderately exposed shores

On limestone ledges on the Isle of Man, clumps of *Fucus vesiculosus* form a complex mosaic together with barnacles, limpets and bare rock (Hartnoll and Hawkins, 1985). Considerable variation in this mosaic was seen over 7 years. *Fucus vesiculosus* varied in total cover from 0 to 65%, the barnacle *Semibalanus balanoides* from 6 to 46%, and the limpet *Patella vulgata* varied in density from 2.5 to 27.5/m². By manipulating the density of various species in small areas, Hartnoll and Hawkins were able to investigate some of the possible causes of these fluctuations. When they reduced limpet density, for instance, a fucoid canopy developed. Although there was no replication here, the result appeared clear cut, and showed the ability of grazers to influence macroalgal settlement. They demonstrated the converse effect of *Fucus* on limpets by comparing the size-frequency of limpets within and outside *Fucus* clumps. Limpets on bare rock were all large (>40 mm long), whereas algal clumps also contained small limpets (<20 mm long), suggesting that *Fucus* clumps might aid limpet recruitment. A complicating factor was added by the effect of the barnacle *Semibalanus balanoides*. Barnacles usually diminish the effectiveness of limpet feeding, and they therefore provide 'refuges' where algal sporelings can 'escape' from grazing. The older the barnacles, the more difficult the limpets find it to graze, and the more escapes of fucoids occur.

We can combine these interactive effects to suggest a mechanism by which the communities on a moderately exposed shore might fluctuate. In brief, the gradual growth of barnacles on the rock surface allows recruitment of *Fucus*. The *Fucus* clumps, as they grow, promote limpet recruitment while discouraging further settlement of barnacles. As the limpet population grows, further settlement of fucoids declines because sporelings are grazed away. The limpets then disperse because the *Fucus* clumps decline, and barnacles can settle again. This cycle can be repeated more or less quickly, or may stabilize at any point if conditions favour one particular state. The model thus provides a series of testable hypotheses which may account for long-term fluctuations seen on moderately exposed shores in north west Europe.

Exposed shores

At Robin Hood's Bay, an exposed shore in north-east England, the dominant organisms are *Mytilus edulis*, *Semibalanus balanoides* and *Patella vulgata* (Lewis, 1977). High on the shore, abundance of all three is fairly constant, as shown by monitoring over a 10 year period. On the mid shore, population densities show more fluctuations, and low on the shore the situation is highly variable. *Mytilus* shows periods of dominance, but predation by *Nucella* and *Asterias* causes it to decline suddenly. Equally important is the loss of *Mytilus* due to wave action, caused particularly when mussels settle densely or in double layers, so that their attachment strength cannot increase in proportion to the surface area facing the waves. Dominance by *Semibalanus* lasts for shorter periods. When *Semibalanus* settles densely, it grows in the so-called 'dog-tooth' form instead of the more normal pyramid shape. In this form, individuals become tall and columnar, and many are literally forced off the rock. As with *Mytilus*, storm action can then erode whole patches. Predation on *Semibalanus* by *Nucella* is sometimes intense.

Although dramatic declines in mussels and barnacles may be caused by predation, we should note that predatory effects are to some extent determined by underlying physical factors. For instance, periodic predation by *Asterias* and *Nucella* is limited to lower shore levels because both predators are intolerant of desiccation. The timing of predation in relation to physical events such as storms is also important in determining whether whole patches of mussels or barnacles are ripped away, and, when they are, what might settle in their place.

Conclusion: the causes of instability

There seems little doubt, from the studies discussed above, that most rocky shores show some degree of instability in their community structure. Equally, there is little doubt that both direct physical influence and the effects of biological interactions are important in determining the changes that occur, and the rates at which individuals and species come to dominate the community. One of the major factors appears to be the influence of some form of 'disturbance' on the community, be it physical or biological. In the next section we discuss the effects of disturbance, and the sequence of events that happens after such disturbance, known as 'succession'.

Disturbance and succession: mosaics on the shore

The major kinds of disturbance so far mentioned have been physical removal of organisms by wave action, and death from predation, itself usually followed by physical removal. There are, however, many physical influences that effectively remove animal communities from the rock surface. In northern latitudes, one of the most important is ice. Many invertebrates can withstand freezing, but moving ice can scour whole communities from the rock. Some mobile animals, such as limpets, are vulnerable to freezing because they can then no longer cling to the substrate. Desiccation may have similar effects, and may, in warmer areas, remove whole stands of algae. Sand scour is another powerful influence, and on exposed rocky shores next to sand beaches the abrasion may be violent.

The most complete studies relating to the effects of disturbance come from the Pacific coast of North America. Here two effects of wave action have been studied. We consider in turn the effects of storms that turn over boulders, and secondly the effects of waves on solid rock shores.

Disturbance on boulder shores

In southern California, much of the rocky shoreline is made up of boulder fields. The boulders are fairly stable in summer, and on their upper surfaces grows a community of algae and invertebrates. In winter, some boulders are turned over by storms, and the community is killed or eaten by grazers if the boulder remains upside-down for long enough. When the boulder is flipped back to its old stable position, the entire surface is bare and recolonization begins again.

Near Santa Barbara, California, boulders low on the shore are normally covered with an algal association dominated by a red species, *Gigartina canaliculata*. On cleared areas, though, bare spaces are colonized within a month not by *Gigartina* but by the ephemeral green alga *Ulva*. By the winter of the first year after clearance, this is replaced by three perennial species of red algae, called the 'mid-successional reds'. Only after 2–3 years does the red *Gigartina* dominate again, covering 60–90% of the area.

To investigate the mechanisms underlying this succession, Sousa (1979) carried out a series of field experiments (Fig. 8.3). First, he cleared *Ulva* from the patches in which it settled. In this case the recruitment of the mid-successional reds was improved. Then he cleared the mid-successional reds, and found that this improved the recruitment of *Gigartina*. So in each case, the early successional algae inhibited invasion by later species.

Fig. 8.3   Growth of *Gigartina canaliculata* (solid circles) and *Ulva* spp. (open circles) in undisturbed plots and in those completely cleared, near Santa Barbara, California. In the undisturbed plots, *Gigartina* maintained nearly 100% cover over 2 years. In the cleared plots, *Ulva* colonized rapidly, but was ousted by *Gigartina* by the end of the second year. Bars show SE. (After Sousa, 1979.)

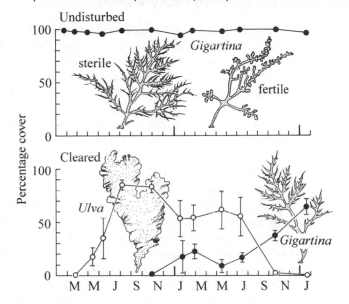

Why, then, do the species replace one another in a succession? Sousa measured the relative susceptibility of the algae to three factors, to see whether long-term resistance to various factors might correlate with later phases in the succession. In terms of desiccation, *Gigartina* is more tolerant than the mid-successional reds, which in turn are more tolerant than *Ulva*. Investigating the palatability to herbivores, he found that the crab *Pachygrapsus*, a major herbivore at the site, prefers *Ulva* to any red algae. By adding *Pachygrapsus* to enclosures in the field he then showed that grazing increases recruitment of *Gigartina*, presumably by preferentially removing *Ulva*. Finally, he looked at susceptibility to the epiphytes that attach to the macroalgae. Several of the mid-successional reds become very heavily overgrown with these epiphytes, and then die leaving space for colonization by *Gigartina*.

In summary, the various phases of the succession are caused by a number of factors. The first phase is colonization by *Ulva* because it produces enormous numbers of propagules, and does not have seasonal reproduction. *Ulva* then resists invasion by the mid-successional reds, but this inhibition is broken by the grazing activity of *Pachygrapsus*, which creates bare space. The mid-successional reds inhibit recruitment by *Gigartina*, but this inhibition in turn is broken by overgrowth of epiphytes, desiccation or herbivory. When *Gigartina* invades, it tends to form a permanent sward because it is long-lived, and it can reproduce vegetatively, occupying any available space around the original plants. In the main, then, the boulders are covered by this species, but disturbance resets the succession to an earlier phase.

Disturbance on exposed rocky shores

The exposed coasts of Washington State are dominated in the mid shore by a mussel, *Mytilus californianus*, which forms large monocultures. Within the mussel beds, however, there are patches that contain macroalgae, barnacles and another mussel, *M. edulis*. On Tatoosh Island, Paine and Levin (1981) investigated the structure of the resulting mosaic by observing succession within these patches, and by examining the rates at which the patches form and disappear. As in the boulder field, most patches are formed in winter—by storm waves, by wave-propelled logs, or by freezing, all of which remove either one mussel or a patch of individuals. In the long run, patches disappear because mussels 'lean in' from the edge, or because some mussels migrate laterally, or because of recruitment of larvae. But in the short term, the sequence of colonization involves settlement by the ephemeral alga *Porphyra* in the first year; then the arrival in the second year of the herbivore-resistant algae *Corallina* and *Halosaccion*, together with some barnacles and another mussel *Mytilus edulis*. Only by the fourth year does *Mytilus californianus* occupy 67% of space, together with a stalked barnacle *Pollicipes* which covers 13%. By year seven, *M. californianus* has reached 79% cover, which it then maintains.

Although the interactions between species were not examined in this example, it is evident that the overall mosaic is caused by frequent patch formation, so that nearby patches are at different stages in a fairly standard succession.

In another example, successions were examined on the Oregon coast. Here the dominant species are a barnacle, *Balanus glandula*, and a fucoid alga, *Pelvetiopsis limitata*. In cleared spaces, the first settlements are made by a small barnacle, *Chthamalus*. Only afterwards do the macroalgae settle, because barnacles decrease the efficiency of grazing by limpets and allow sporelings to attach. *Balanus* finally outcompetes *Chthamalus*, to recreate the climax of the succession. Although the species concerned are very different, there are many parallels here with the example from boulder shores.

The mosaic-cycle concept

There are now many examples to show that disturbance, either physical or biological, may reset natural successions. In California, for instance, subtidal kelp communities dominated by *Macrocystis* occur in patches. Since these natural successions eventually drift out of phase even when starting at a 'virgin' site, the result is a mosaic of the type discussed above.

Remmert (1991) took the idea of mosaic communities one step further, and suggested that in fact most ecosystems, or patches within them, should be regarded as cyclical. The 'mosaic-cycle concept' that he presented thus places a slightly different emphasis on the system: as individuals grow older, they themselves bring about physical disturbance because with their larger size and greater resistance they are more vulnerable to physical disturbance, or alternatively present better targets for predators. This mosaic-cycle concept has been applied particularly to terrestrial forests, but fits well with many rocky shore ecosystems.

Community structure: competition as a structuring force

In the previous section, we concentrated mainly upon the effects of physical disturbance on community structure. The studies discussed have, however, made it quite clear that interspecies interactions are crucial in determining the course of successions and thus in structuring many communities. In this and the next two sections, we discuss some of the evidence regarding the importance of competition, predation and grazing.

Interspecific competition

We have already described some convincing experiments showing the effect of interspecific competition between algal species (p. 24). The result of this on north European shores is to produce a strict zonation pattern: the faster-growing species, which are not desiccation-tolerant, live downshore from the slower-growing ones, which can tolerate greater desiccation extremes.

The classic account of the influence of competition in rocky-shore communities, however, concerns competition for space between two barnacle species (Connell, 1961a). On an exposed shore, on the Isle of Cumbrae in Scotland, Connell investigated competition between *Chthamalus stellatus* and *Semibalanus balanoides*. Where the two species settle together, between MTL and MHWS, considerable crowding occurs. To investigate the possible effect of *Semibalanus* on *Chthamalus*, Connell removed all the young *Semibalanus* that were touching or surrounding *Chthamalus*, in several plots. This was carried out after the period of settlement was over. Meanwhile, he marked out adjacent control (untouched) areas. Both experimental and control plots were set up at several different tidal levels. Monitoring over the next year showed that between MTL and MHWN the survival of *Chthamalus* is very much better where *Semibalanus* has been removed than where it is allowed to grow (Fig. 8.4). Direct observation showed that growth of *Semibalanus* is often an important cause of death in *Chthamalus*: many of the *Chthamalus* are killed when *Semibalanus* grows over and smothers them; while others are lifted off the rock when *Semibalanus* undercuts them. In other cases *Semibalanus* appears to kill the *Chthamalus* when it grows tall and crushes them laterally.

Fig. 8.4   The effect of the barnacle *Semibalanus balanoides* on the survival of *Chthamalus* on the Isle of Cumbrae, Scotland. At high tidal levels (1 m above MTL, approximately MHWN), no competition is observed. Near MTL, however, *Semibalanus* outcompetes *Chthamalus* if not removed experimentally. (After Connell, 1961a).

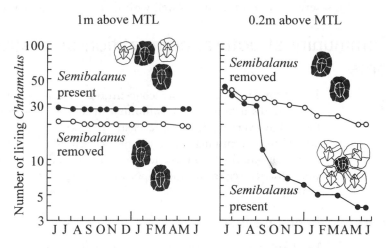

In contrast to these experiments between MTL and MHWN, those carried out above MHWN showed no effect of *Semibalanus* on *Chthamalus*. Competitive effects of *Semibalanus* are therefore greater at lower tidal levels, and this competition for space is a major factor in restricting *Chthamalus* to higher tidal levels. In fact, *Chthamalus* normally survives as adults only above MHWN.

Connell examined the competitive effect of *Semibalanus* on *Chthamalus* at lower levels on the shore by transplanting rocks from high levels. Here the effectiveness of *Semibalanus* as a competitor is reduced because the dog whelk *Nucella* is common, and it reduces the numbers of *Semibalanus* significantly.

Evidently it is important not to consider the effects of competition in isolation, but to take into account the indirect effects of predation and other physical and biological factors. On top of these complications, it should be noted that almost all experiments carried out in the field introduce some unnatural conditions. Although such experiments may demonstrate that competition *can* occur, they do not show that it necessarily *does* occur in normal situations. Predation, for instance, may keep population densities down to levels at which competition becomes insignificant.

Nevertheless, many experiments have demonstrated that competition can occur on the shore (Paine, 1994). A large proportion of these concerns not the competition for space by sessile animals, but competition for food by mobile grazers. In New South Wales, Australia, for example, three species of limpets forage together for food: a patellid, *Cellana tramoserica*, and two *Siphonaria* species, which are pulmonates. *Cellana* grazes on microalgae and very small sporelings, while the *Siphonaria* species graze on macroalgae, leaving behind the basal parts of algal thalli. Do these species compete? To investigate this possibility, Creese and Underwood (1982) placed limpets in fenced areas at a variety of densities and species combinations, and then monitored their survival over 12 weeks. The *Siphonaria* species showed only low mortality when they were enclosed together, even at high densities. But when enclosed with small numbers of *Cellana*, their mortality rose. This is clear evidence that *Cellana* has a direct competitive effect on *Siphonaria*. The explanation probably lies in the difference in the modes of feeding of the two types of limpets. Because *Cellana* scrapes off microalgae, it removes not only diatoms but small sporelings of macroalgae. *Siphonaria* species are dependent upon macroalgae, so that when *Cellana* decreases macroalgal recruitment, they starve. There is no converse effect of *Siphonaria* on *Cellana*, because *Siphonaria* always leaves behind a film of algae and basal thalli, which can be utilized by *Cellana*.

Intraspecific competition and facilitation

Competitive effects within species have not often been clearly differentiated from the interspecific effects described in the previous section. We take just one example to show that intraspecific competition can be important. The grazing gastropod *Nerita atramentosa* is abundant on sheltered rock platforms in New South Wales, Australia. By enclosing different densities of marked snails within mesh cages, Underwood (1976) could measure both overall mortality and the growth rates of individual juveniles. At normal densities, juveniles grow rapidly, but at five times the normal density they show no growth. Adults, in contrast, show no growth at any density, and at high densities they lose tissue weight.

In these experiments, the relation of percentage mortality to density also differs between juveniles and adults (Fig. 8.5). Juveniles show very low mortality ($<10\%$) even at five times normal density. Adults, on the other hand, show high mortality ($>50\%$) when density is five times normal. This effect is the same whether high densities are created by adding juveniles or adults. Since increased densities in the experiments produce loss of tissue and subsequent death in adults, but no deaths in juveniles, competition for food certainly could regulate population density in *Nerita*: at high densities, juveniles outcompete the adults for food, and this could lead to regulation of the population density if, say, there were large-scale juvenile recruitment.

Fig. 8.5 Mortality of various densities of the gastropod *Nerita atramentosa* in Botany Bay, Australia. Snails were confined in mesh cages of 23 x 23 cm. Adult mortality rose greatly with increasing density, while juvenile mortality did not. Bars show 95% confidence limits. (After Underwood, 1976.)

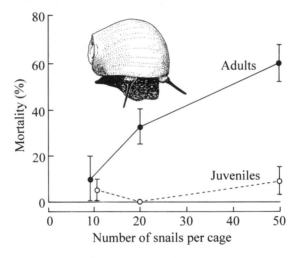

In this kind of situation, it is always possible that some other factor would in fact keep population densities below levels at which competition occurs. For *Nerita*, this seems unlikely because predation levels, at least, are low.

The opposite effect to that of competition is called facilitation: here individuals actually increase the survival of their neighbours. An example is provided by the barnacle *Semibalanus balanoides* in New England (Bertness, 1989). Here the survival of individuals high up on the shore is *increased* when they are crowded together: crowding reduces the temperature of both the rock and the barnacles, and this increases survival in summer.

Aggression

One of the interesting questions about competition that we have not so far discussed is how competition is manifested. What kind of interactions between individuals actually occur? In the case of sessile species such as barnacles and bryozoans, growth and smothering seems to be the most common interaction. In the grazing *Nerita*, there appears to be no actual physical interaction: competition occurs via utilization of food resources. But in some other species, individuals interact and show aggressive behaviour.

We have already discussed one example of this in the section on sea anemones (p. 129). In that case, aggression appears to give individuals or clones their own 'personal space', and this may be advantageous if there is a limited food supply. Here we discuss another example, this time concerning limpets.

On the Californian coast, the largest limpet is an acmaeid called *Lottia gigantea*. This species not only grazes on microalgal films, but each individual actively maintains a territory around itself, which is visible as a film of algae with radula marks (Stimson, 1970). Within this territory, each limpet has a home scar, and returns to it after grazing. Outside the grazing areas, several other kinds of limpets (*Acmaea* species) are common. Why are *Acmaea* species never found within the *Lottia* territories? If an individual of *Acmaea* is placed near an individual of *Lottia* as the tide starts to cover the area, the results are startling. As the *Lottia* moves forward, it contacts the *Acmaea*. It stops, then lowers the forward edge of its shell, and moves its foot forward, striking the shell of the *Acmaea* with its own. It continues to shove the *Acmaea* until the 'intruder' has been pushed out of its territory. These 'staged' encounters are backed up by observation of a few natural ones: any grazing intruders are pushed out by *Lottia*.

Similar behaviour has also been reported in patellid limpets on the shores of southern Africa. Here species such as *Patella longicosta* occupy territories or 'gardens' in which grows the alga *Ralfsia*. These gardens are defended against invaders.

Conclusion: how important is competition?

Both intra- and interspecific competition are common features in the rocky intertidal zone. Nevertheless, it has been suggested that the importance of competitive interactions may very often be reduced because predation or disturbance or some other factor may reduce population densities of potentially competing species to such low levels that there is sufficient of the appropriate resources for all species. On this 'intermediate disturbance hypothesis', competition is important only in moderately harsh conditions, and not where physical conditions are very harsh or very benign. In very harsh conditions, populations are reduced directly by physical stress, while in benign conditions they are reduced by predation. There are, however, several other theories about the situations in which competition may be important in structuring the community. It has been suggested, for instance, that plants and carnivores are more likely to be regulated by competition than herbivores. Another possibility is that higher trophic levels should be more affected by competition than lower ones. But regardless of which theory is advocated, levels of competition may be regulated by further factors such as recruitment. Especially in those species that have pelagic larvae and settle in extraordinarily high densities, such as barnacles and mussels, the blanket coverage of spat may produce such close packing of individuals that there is almost certain to be a high level of competition.

There has therefore been much argument between the proponents of the various theories, but as yet no general consensus. Particular instances can be cited to support each theory, but different situations seem to provide different answers. At least one important point can, however, be made concerning the types of competition involved. In such examples as competition between barnacles, the resource (i.e. space) cannot be partitioned between species. Co-existence of species, and maintenance of community structure is, therefore, heavily dependent upon larval settlement which replaces those individuals that are lost. In contrast, in the example of competition between grazers, the resource (i.e. algal food) *can* be partitioned between species so that two (or more) species can co-exist. It is not, perhaps, surprising that one theory cannot cover two such different situations.

Community structure: predation as a structuring force

In previous chapters, several suggestions have been made concerning the importance of predators in the biology of their prey: sea otters and fish consume sea urchins; crabs consume mussels; nudibranchs consume bryozoans. To these examples we should add the effects of parasites and disease, such as the amoeba that attacks sea urchins (p. 101). So little is known about diseases of marine invertebrates, however, that we consider here only the evidence concerning predators. How important are predator attacks? Do they affect the population densities of prey species ? Do they affect the community as a whole?

Cases in which predation is a dominant force

On San Juan Island, Washington, predatory whelks (species of *Thais*) feed on the barnacle *Balanus glandula*. In exposed situations, the barnacle survives to maturity only at the top of the shore; but in sheltered bays and on isolated pilings, adults occur widely. Since predators are absent in the sheltered bays, but common on exposed coasts, might it be that predators eliminate the barnacles wherever they have access to them? To test this idea, Connell (1970) excluded predators using stainless-steel mesh cages. Inside these cages, *Balanus* lived for several years, while at adjacent control sites each annual settlement was killed (Fig. 8.6). The natural survival of *Balanus* at the top of the shore thus appears to be due to inaccessibility to *Thais*. In this example, the barnacle mortality ascribed to predation is a major factor, while in the situation in Scotland described on p. 167, inter- and intraspecific competition are more important than predation. The difference between the two sites may be related to the regularity of barnacle recruitment: in Scotland recruitment is variable, whereas in Washington it is very regular, and has allowed the evolution of a specialized predator, one of the *Thais* species. If this is so, the importance of considering the interrelated effects of competition, predation and recruitment is evident.

Fig. 8.6   The effect of the whelk *Thais emarginata* on the survival of the barnacle, *Balanus glandula*, near Friday Harbor Laboratories, Washington State. Barnacles protected by cages live many more years than those subjected to predation. (After Connell, 1970.)

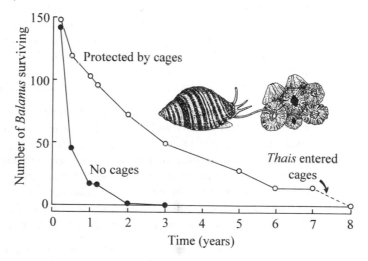

At Mukkaw Bay, on the exposed coast of Washington, the mid shore supports a varied community. The herbivores and suspension feeders comprise two chitons, two limpets, the bivalve *Mytilus californianus*, three species of acorn barnacle and the stalked barnacle, *Pollicipes polymerus*. A predatory whelk, *Thais*, feeds on the acorn barnacles and *Mytilus*, and a predatory starfish, *Pisaster ochraceus*, feeds on all these species, including *Thais*. The total community on horizontal surfaces contains many more species than those already mentioned. Paine (1974) described the effects of removing the 'top predator', *Pisaster*. This could not be done using cages because of the large wave forces, so starfish were removed manually. Over 3 years, the result was that more than 25 species disappeared from the area, which became covered with *Mytilus californianus* and some *Pollicipes polymerus*. The community changed from one of high diversity to one of low diversity.

Paine's explanation for this result was that *Mytilus californianus* is a 'competitively dominant' species: it can exclude almost all other species from 'primary space' (i.e. bare rock). The normal action of *Pisaster* when present is to eliminate *Mytilus*, thus allowing a wide variety of other invertebrates to utilize the primary space: the barnacles, chitons and limpets listed above, as well as many species of sponges and encrusting algae. In this case, *Pisaster* acts similarly to the presence of physical disturbance: both processes free primary space for settlement. Similar starfish–mussel interactions can have important effects on the shores of New Zealand and Chile. Paine's hypothesis that predators can have overwhelming effects on the shore—the 'keystone species hypothesis'—is illustrated in Fig. 8.10, and we discuss it further on pp. 183–5.

Cases in which predation is not the dominant force

Evidently, predators seldom eliminate their prey entirely, so there must be situations in which their actions are not so important, either for the individual prey or for the community. As an example, we may note that some attempts to repeat Paine's experiments in which *Pisaster* is excluded have not resulted in subsequent dominance by *Mytilus*. This may have been because in these later years, *Mytilus* recruitment was poor—thus reinforcing the importance of recruitment (Paine, 1974).

There may be a number of situations in which predators are not effective in reducing prey density and thus in eliminating competition. First, the level of recruitment of prey is important. If this is particularly high, population density may remain high long enough for competition to occur, before predators have any effect. Thus settlement by barnacles can be so dense that the predators (*Thais* species) cannot consume a significant fraction. In the contrasting case, if recruitment is very low, there may be no competition for predators to interfere with.

Secondly, the level of predation may be reduced because of some physiological limitation of the predator, or the effect of some higher predator. On the Washington coast, *Pisaster* cannot attack *Mytilus* above a certain tidal height because of its lack of desiccation tolerance.

Thirdly, prey may reach some kind of 'refuge' from predators. If, for example, conditions for prey are favourable, they may grow quickly so that they reach a refuge in size. For example, *Mytilus californianus* on the Washington coast that reach 3 cm in length are not attacked by *Thais* (Dayton, 1971). There may also be physical refuges from predators such as the pits or crannies in the rock where large predators cannot reach, and mobile grazers can shelter. Alternatively, the predators themselves may utilize crevices at low tide, so that prey more than a certain distance away from a crevice will be inaccessible to them. Fairweather (1988), working in New South Wales, showed that there were haloes bare of prey around crevices. The crevices shelter predatory whelks such as *Morula marginalba* at low tide, but these emerge under water, and sometimes form feeding 'fronts' at the edge of dense patches of barnacles. When the whelks are manually removed, the 'haloes' disappear.

Fourthly, predators can be resisted to some extent. Many prey have anti-predator structures, chemicals or behaviour patterns. Some of these prey species may even depend for their existence upon a predator: if they are competitively inferior to others in a similar guild, but can resist the predator better than their competitors, they will survive only where the predator is active.

Lastly, many prey species are clonal organisms. These are unlikely to be eliminated by predators, since some of the clone is likely to survive any predator attack. Bryozoans, for instance, are rarely totally killed by nudibranchs. Instead, the predators open a clear patch within the clone, and here regeneration or recruitment may stimulate further competition. This is an example of partial predation.

In summary, predation can be the dominant force structuring communities, but it is not necessarily so. The influence of predation must be integrated with competition, recruitment and the effects of physical forces when attempting to construct any model for community structure.

Community structure: grazing as a structuring force

Because grazing has many similarities to predation in its community effects, some authors have included it under the heading of predation. Grazing is, however, a very different process and involves totally different species from predation, so here we keep to the conventional separation. We have already discussed some of the effects of grazers on microalgae (p. 75) and macroalgae (p. 81). How do these effects rank with, and interact with, those of competition, predation and physical disturbance? Two examples show some of the effects of herbivory on intertidal communities.

On the low shore of Washington coasts, dominance by the brown alga *Hedophyllum sessile* varies enormously. Is this variation related to the abundance of grazers? Dayton (1975) removed *Hedophyllum* from areas where it was abundant and followed subsequent events. Even when molluscan herbivores (a suite of gastropods and chitons) were allowed to graze, the alga recolonized the area rapidly. Physical removal of the major molluscan grazer, the chiton *Katharina tunicata*, had no effect on algal regrowth. It appears that these herbivores have no significant effect on the composition of the algal community, probably because predators such as *Pisaster* and others keep the grazer populations at a density well below the carrying capacity of the environment.

However, when Dayton removed the urchin *Strongylocentrotus purpuratus* from a strip of shore on which there was no macroalgal growth, algae bloomed dramatically (Fig. 8.7). First, a wide variety of fugitive species such as *Ulva* and *Porphyra* dominated the strips, and after 2 years *Hedophyllum* reached up to 40% cover. The influence of *Strongylocentrotus* on the algal community is therefore much greater than that of the molluscs. *Strongylocentrotus* itself appears to be controlled by predatory starfish, and wherever these reduce the urchin populations, patches of algae become established. In this situation, grazing by urchins is therefore in some places a dominant feature, controlling the algal community, but in others its effects are removed by the actions of a predator.

Fig. 8.7   The effect of removing the grazing sea urchin, *Strongylocentrotus purpuratus*, on algal growth at Shi Shi, a shallow reef in Washington State. Fugitive species such as *Porphyra* colonized rapidly, but *Hedophyllum sessile* and understorey species such as *Corallina* did not arrive until several years had passed. (After Dayton, 1975.)

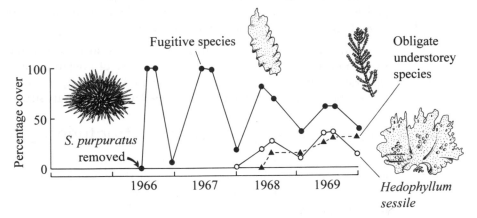

In comparable situations in New England, the rock is dominated by the mussel *Mytilus edulis* on exposed headlands, but by the red alga *Chondrus crispus* in shelter. The major herbivore on sheltered shores is the winkle, *Littorina littorea*, and although this has little direct effect on *Chondrus*, it consumes ephemerals such as *Ulva* and *Enteromorpha* (Lubchenco and Menge, 1978). Since these ephemerals normally suppress the growth of *Chondrus*, the presence of *Littorina* increases the rate of settlement and growth of *Chondrus*.

The grazer here is therefore important in determining algal community structure. However, the presence or absence of *Chondrus* is further determined by its competition with the mussel *Mytilus*. At exposed sites, *Mytilus* outcompetes *Chondrus*. At sheltered sites, *Mytilus* does not outcompete *Chondrus* unless its predators are excluded by experimental cages: normally, these predators eliminate *Mytilus* almost completely. The situation at exposed sites, where *Mytilus* is dominant, is evidently due to a lack of predators. In this example then, the interlocking themes of competition, predation and grazing are all important, and relate also to the degree of physical disturbance caused by wave exposure.

The influence of recruitment: 'supply-side' ecology

Some of the species living on rocky shores produce their offspring as 'miniature adults'. For example, many whelks and some winkles do this, so that adult populations are wholly governed by conditions on the shore. The very great majority, however, including algae, limpets, mussels, barnacles, crabs and starfish, have pelagic larvae which float in the plankton before settling and metamorphosing into adult form. For these organisms, it has long been evident that recruitment from the pelagic phase must be outstandingly important in governing the density of populations on the shore. Recently, the factors determining this recruitment or 'supply' of propagules have been gathered under the name 'supply-side' ecology.

Larval biology

The majority of invertebrates which produce planktonic larvae invest very little energy or material in each propagule. The propagules develop their own feeding mechanisms and grow using the energy obtained from smaller plankton, much of it often phytoplankton. Such larvae are called 'planktotrophic'. A minority of invertebrates, such as some of the nudibranchs, endow the larvae with their own food supply in the form of yolk, so that they do not have to rely upon external planktonic food. These larvae are called 'lecithotrophic'. For algae, the situation is slightly different, in that the spores do not usually develop during the planktonic phase, and therefore do not require external nutrients until they are about to settle. In terms of their dependence upon the planktonic phase they have some similarities with lecithotrophic larvae, since this stage acts purely for dispersal.

Two aspects of larval biology are particularly important in respect of the community structure to be found on the shore. First, there is the complex of factors determining larval survival. Is there enough planktonic food to allow larvae to grow? Can the larvae maintain their appropriate position in the water column? Can they avoid being eaten by predators? Secondly, the larvae must be able to settle and change into adults. How do they choose appropriate sites? How successful is metamorphosis? These questions are, at least in detail, really beyond the scope of this book, but must be borne in mind when considering recruitment. In general, we can be sure that losses of larvae are enormous. Many sessile invertebrates produce thousands of planktotrophic larvae, yet only a few will grow to maturity. Any species with this type of development depends heavily upon conditions in the water column, and the timing of egg/larval release in relation to the spring bloom of phytoplankton may be crucial for them. Lecithotrophic species produce far fewer larvae, but for them the escape of each from predation becomes more crucial.

Many of the critical details of larval life in the water column are so far unknown. Some species disperse great distances offshore, so that larvae colonizing a particular area may have come from remote populations. The larvae of the brittlestar, *Ophiothrix fragilis*, for instance, can be found as far as 50 km offshore in the Mediterranean (Fig. 8.8). The times of appearance of these larvae and those of sea urchins correspond with spawning periods of the adults, suggesting that there is considerable synchronization of spawning between populations. Spawning must therefore produce short but massive release of gametes (Pedrotti, 1993). Disappearance of the larvae of the sea urchin *Paracentrotus lividus* from the plankton in late autumn coincides with the observation of newly settled urchins (< 1 mm in diameter) on benthic substrates. These types of observation suggest that over a number of years, it should be possible to make an assessment of the importance to recruitment of variation in larval production, survival and dispersal.

Fig. 8.8   The abundance of different larval stages of the brittlestar, *Ophiothrix fragilis*, at 10 m depth on a transect offshore from Nice, southern France. The hauls were made on 30 October. Categories are stacked above each other so that the upper boundary shows total larvae. (After Pedrotti, 1993.)

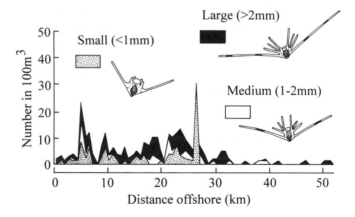

The study of larval settlement has advanced further than study of life in the plankton. Some of the pecularities of larval settlement have already been discussed in Chapter 6. In barnacle species such as *Chthamalus stellatus*, for instance, the vertical distribution of adults on the shore is partly determined by larval choice: larvae settle densely only above MTL, while those of *Semibalanus balanoides* settle all over the shore. Since larvae of *Semibalanus* also choose to settle where there are already members of their own species, this behaviour to some extent favours crowding, and promotes intraspecific competition. Nevertheless, the settlement behaviour of very few types of propagule has been investigated in detail, and there has been much discussion about the possibility that the majority undergo 'passive deposition'—in other words that they behave much like inert sediment particles until they actually attach.

In particular it has been suggested that the propagules of some of the algae settle by passive deposition, since most are not motile. Examination of the spores of *Sargassum muticum*, a laminarian alga, however, suggest that this is not so (Norton and Fetter, 1981). The spores of this species settle at less than 1 mm s^{-1} in still water, so one would expect settlement to be slow. But when they are allowed to settle in flowing sea water within the laboratory, spores settle within a 30 s period. How is this achieved? Like many spores, those of *Sargassum* are sticky, and when the turbulence of the water flow flings the spores against the substrate, they stick to it. The spores therefore reach the substrate by a process of turbulent deposition.

In contrast, the motile swarmers of green algae seek out crevices in which to settle, and there is probably a great range of responses to water currents and substrates among various planktonic propagules. This variety is exemplified by the larvae of invertebrates whose adults live in soft sediments: some can make choices of sediment type in which to settle when they are in a current, while others cannot. The details of settlement of propagules would appear to be a fruitful area for study.

Variations in recruitment on the shore

A crucial question to ask about larval 'supply' is how much the variations in recruitment really affect populations on the shore. In Monterey Bay, California, Gaines and Roughgarden (1985) were able to examine this problem in a high-shore barnacle, *Balanus glandula*. They chose two sites, only 30 m apart, one with low settlement rates, and the other with rates 20 times higher.

Where settlement rates are low, there is a great deal of variation in adult abundance *from year to year*, although numbers change little *within each year*. At this site, recruitment probably fluctuates because of the variation in survival of the larvae in the plankton, as well as variation in the currents that bring the larvae to shore. Once settled, densities are low and there is little interaction between individuals. The situation is different where settlement is high. Here there is little variation in adult abundance from year to year, but the densities change drastically throughout each year. At this site, there are apparently always enough larvae to saturate the system, and high population densities lead to great intraspecific competition. Mortality is also high, due to predation by *Pisaster*, which is presumably attracted by the dense food supply.

The overall effects of differential recruitment at the two sites may thus be of equal importance to the factors that affect the populations once they have settled. If recruitment is high, subsequent biological interactions may be important; but if recruitment is low, the situation may parallel that in which predation, wave-disturbance or grazing lowers population densities, and biological interactions will then have much less significant effects.

Experimental investigations: how to obtain evidence about the factors structuring communities

From the studies so far discussed in this chapter, it must be apparent that there is no simple answer to the question 'What governs community structure on the shore?' Organisms may compete for space or food, and this may lead to some form of succession. In some cases predation or grazing will reduce population densities below those at which competition is significant. At times, physical disturbance may remove dominant competitors, providing new primary space and allowing the succession to restart. On top of these processes, the variability in recruitment may dictate changes in community structure, and may affect the importance of competition and predation. In any one situation, it is thus unlikely that an examination of one process or interaction will give a very broad insight into community structure. The question then arises, of the way in which meaningful investigations can, or should, be made.

This question can finally be resolved into one of experimental design, and it is appropriate to discuss this briefly before considering theories that aspire to an explanation of the structure of communities. It is essential that the basic principles are understood, because to some extent the design of the experiments dictates the answers received.

Experimental manipulations can be very useful in testing hypotheses concerning both influences of physical factors and interactions between species. Indeed, several of these have been discussed in the foregoing sections. However, the hypotheses to be tested must be developed logically, and the experiments themselves must be designed carefully and must contain appropriate controls. In particular, hypotheses should be formulated to allow the testing of a variety of alternative models of the system, and not just one model. As an example, we may take tests for four hypotheses concerning the factors that determine the distribution of macroalgae in New South Wales (Underwood, 1985).

The low shore in New South Wales is dominated by foliose macroalgae, and the algal belt has a relatively abrupt upper limit. Underwood considered four alternative models that might explain this distribution, and then went on to make hypotheses predicting what would happen if any or all of these were correct. This allowed him to set up experiments to test the four hypotheses experimentally. However, such experiments must be set up to test *null* hypotheses, these being phrased to suggest that the results of any experiment will be *other than* those predicted by the original hypotheses. By testing the null hypothesis and not the original hypothesis, it is possible to make tests of falsification, i.e. if the null hypothesis is disproved, this lends support to the original hypothesis. This seems a very round-about procedure, but it is necessary because, as has been known for centuries, it is almost impossible to prove a hypothesis correct, while it is possible to disprove one.

The four alternative models considered separate possible factors that might control the foliose macroalgae high on the shore:

(1) their inability to withstand high temperatures and desiccation;
(2) their susceptibility to grazing;
(3) their ineffective recruitment; and
(4) their inability to compete with encrusting algae.

Underwood formulated hypotheses based on each of these four models, and then constructed null hypotheses to include alternatives to them. Thus he proposed to test the following null hypotheses:

(1) that desiccation might have *no* effect on algal distribution;
(2) that grazers did *not* prevent upward distribution of algae;
(3) that recruitment did *not* limit algal distribution; and
(4) that encrusting algae did *not* outcompete foliose forms.

For the tests, he set up experiments using control areas, mesh fences, cages and roofs (Fig. 8.9) to manipulate the variables under test. Control

Fig. 8.9 A suggested experimental design to investigate the effects of physical stress and grazing on the distribution of low-shore algae. Comparisons of the various treatments allow some estimate of the importance of shading and of grazing. Note, however, that effects of the treatments on shading are all slightly different. (After Underwood, 1985.)

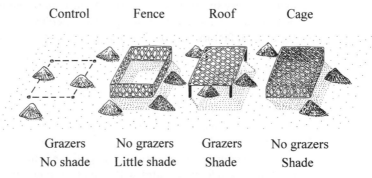

Control	Fence	Roof	Cage
Grazers	No grazers	Grazers	No grazers
No shade	Little shade	Shade	Shade

areas do not affect the degree of desiccation, or access by the grazers. Fences exclude grazers but scarcely affect desiccation. Cages exclude grazers but reduce the degree of desiccation, while roofs (without sides) allow the grazers in but also reduce desiccation. Underwood also used replicate sets of experiments, in which the encrusting algae were removed, testing for effects of competition. The results were clear. Wherever the grazers were excluded (fences and cages), foliose macroalgae colonized the rock surface, and there was little effect attributable to the presence of encrusting algae. The null hypotheses concerning recruitment and competition (3 and 4) can therefore not be rejected, and the last two hypotheses are eliminated: algal recruitment is not limiting, and neither is competition from encrusting algae.

In the fenced plots, algae colonized but did not grow to adult size. Only in the cages did algae colonize and grow into adults. Null hypotheses for hypotheses concerning desiccation and grazing (1 and 2) can therefore be rejected: both grazing and desiccation must be accepted as factors controlling macroalgal distribution. Grazing limits upward distribution of the plants, but some physical factors such as desiccation and/or temperature also influence growth, and the two factors must be considered together.

This example raises new problems that relate to yet more interactions on the shore. For instance, if the grazers are a major factor in limiting upshore distribution of macroalgae, why are they not more effective at lower levels? Is this to do with their physiological intolerance of, say, submersion, or are they, perhaps, limited by low-shore predators? These, and other models, can be tested as alternatives using the same protocol of hypothesis, null hypothesis and experiment.

The experimental investigation of rocky shores can be seen, from this brief outline, to be far from a simple matter. There are many other complexities, as well as alternative views of how experiments should be carried out. Here we mention two additional points. First, there is a need for adequate replication, avoiding the perils of 'pseudoreplication'. The latter problem arises when experimenters carry out replicate experiments in very small areas of the shore: the experimental results may then be typical only for that very small area, whereas some distance away the controlling factors may be different. Experiments should therefore be set up in several separate plots, each of which has experimental and control areas within it.

A second point has been made by Paine (1994). He suggested that the use of unmodified 'control' areas as reference points in experiments is not the most effective way forward, because these unmodified areas are very variable and subject to many unknown influences. Instead, he prefers to use 'monoculture' areas, produced by the elimination of a grazer or a predator, as reference states with which to compare other treatments.

Given the complexity of ecological situations on the shore and the necessity for complex experimental set-ups in order to test hypotheses, it is perhaps not surprising that there has been much disagreement about the major factors structuring communities. We now summarize this discussion.

The keystone species hypothesis: does one consumer control the community?

In some of the examples of predation given above, the action of the predator appears to control the abundance of its prey. When the predator (such as a starfish on the Californian coast) is present, mussels are infrequent or absent. In the absence of mussels, many other species such as limpets and sponges flourish. When the starfish are removed, the mussels flourish and the limpets and sponges disappear. From observations such as this has emerged the concept of 'keystone species': species such as the starfish, which are essentially the major factor controlling community composition and diversity. More correctly, the community may be postulated to depend upon the *interaction* between such a consumer and its major food resource. Another example might be the limpet–microalgae interactions in north-west Europe: when limpets graze algae away, a barnacle-dominated community develops; when limpets are not present, macroalgae come to dominate the shore. We discuss in turn the suggestion that such keystone species are of overriding importance, and some counter-suggestions that they are rarely of widespread influence.

Keystone species? The community structure of kelp beds

If keystone species do exist, how do the links within a system dominated by such a species actually function? Paine (1980) proposed that the trophic link between a top consumer and its resource would be a 'strong' one, while those between other consumers and their food might be 'weak'. Only the 'strong' links would be important in determining community structure, but on these strong links might depend whole 'modules' of weakly interacting species that flourish or disappear depending upon the presence or absence of the strong link (Fig. 8.10). In the example of starfish and mussels, starfish would be the top consumer, the mussel its major resource, and the whole network of grazers and suspension feeders that appear when the starfish is present would be the 'module'.

Paine used another example to demonstrate the keystone species hypothesis, and this has been much discussed. In the beds of giant kelp, *Macrocystis*, off the Pacific coast of North America, there used to be large numbers of sea otters (*Enhydra lutris*). This species was decimated in the nineteenth century by hunting, and has only recently been re-established now that hunting has been banned. One of the major food items of the sea otter is a sea urchin, *Strongylocentrotus*, and it has been hypothesized that the decline in sea otters allowed the sea urchins to survive in greater numbers than before. The sea urchins are grazers, and Paine attributed to their increased grazing pressure the decline of algal abundance on the California coast. More than this, some authors have suggested that, since urchins are capable of deforesting some areas, providing an extreme contrast to areas of dense kelp where urchin densities are low, the urchins trigger a switch from one 'stable state' to another (p. 140). Within their geographical range, the otters may mediate these switches through their predation on the urchins. Here, then, is a possible hierarchy of strong interactions, sometimes called a 'cascade' effect.

Fig. 8.10   A model of interactions between species on the shore. C represents consumers; R shows resources. The thickness of the arrows shows strength of interaction. In this case C_1 is the 'keystone species': its removal results in an increase in R_1 and consequent decrease in R_2, R_3, C_2 and C_3. Removal of C_2 or C_3, however, has little effect on the system. The model is illustrated with reference to the experiments of Paine (1974) at Mukkaw Bay, Washington State, where the 'module' in fact consists of at least 20 species in addition to those shown. (After Paine, 1980.)

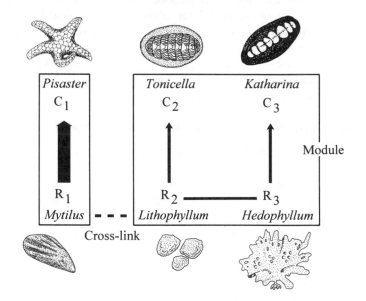

Discussion and testing of the hypothesis that sea otters are keystone species in kelp beds has been intense: are they keystone species or are they 'just another brick in the wall'? One line of evidence comes from following changes in the kelp canopy after sea otters reoccupied areas in which they had long been extinct. At Monterey, for instance, the sea bed had high densities of sea urchins in the late 1950s. Ten years after sea otters reached the area, the urchins were confined to crevices, and giant kelps were abundant. It certainly seems that otters *can* have a striking effect on the ecosystem. It is, however, important to know how widespread this effect actually is. Surveys of shallow subtidal reefs outside the present range of the otters, but within their ancestral range, have found only a few sites with 'urchin deforestation' (equivalent to the 'barrens'

described on p. 85)—about 9% of the total (Foster and Schiel, 1988). Only in this small fraction of sites could sea otters therefore act as keystone species if they returned. At most sites (89%), the urchins are present in patches, or are restricted to crevices, while only 2% are completely dominated by algae. Many sites changed their characteristics over the survey period, suggesting that rather than switching between 'stable states', the sites show intrinsic variability in composition. Kelp communities may not, therefore, generally be structured around inter-actions between sea otters, sea urchins and macroalgae. Instead, varia-tion in the communities may depend upon a whole suite of factors, the 'bricks in the wall': the availability of algal propagules; the nature of the substrate, nutrient and light levels; incidence of storm action; as well as the effects of otters and urchins. The 'hierarchical' view that sea otters ameliorate the grazing effects of sea urchins is appealing because of its simplicity, but it does not take into account these other factors.

There have been many other comments on the system. Tests to mimic the action of sea otters by manually removing numbers of urchins have failed to produce large changes in kelp density. In general, the otter–urchin interaction appears to be important in some areas, but not in others, and it has been suggested that it is not an effective factor over very wide areas. But this leads to the question of how large an area does a species have to dominate in order to be called a keystone species? The scale of experiments has, by definition, been quite small, yet the concept is applied to large but undefined areas. In the long run then, the question of whether species can be called 'keystone species' or not may be a question of scale.

Multi-species interactions: community structure of rocky shores in Australia

On the mid shore in New South Wales, the dominant organisms in regions of moderate exposure are barnacles, limpets and whelks. Under-wood *et al.* (1983) investigated the roles of the four most common species in this community by means of experimental manipulations. These species are the barnacle, *Tesseropora rosea*; the large patellid limpet, *Cellana tramoserica*, which feeds on open rock surfaces; the small acmaeid limpet *Patelloida latistrigata*, which mostly grazes in the interstices between barnacles; and the predatory whelk, *Morula marginalba*. The experiments of Underwood *et al.* differ somewhat from many that have been discussed so far, in that they examined interactions by varying the *densities* of species, instead of simply removing them. For example, to study the effect of the limpet *Cellana* on the survival of barnacles, they excluded the whelk *Morula* and confined *Cellana* at varying densities per enclosure (Fig. 8.11). Over 10 months, barnacles showed high survival where there were two or four *Cellana* per enclosure, but low survival where there were no *Cellana* or where *Cellana* were at high densities. At low *Cellana* densities, macroalgae grow and choke the barnacles, while at high *Cellana* densities the limpets crush the newly settled cyprids and young barnacles. At intermediate densities, the limpets graze down the macroalgae sufficiently to encourage barnacle settlement and growth. In parallel experiments, where whelks were allowed access, barnacle survival was low at all *Cellana* densities—predation here takes over as a dominant factor.

Fig. 8.11   Interactions between the barnacle *Tesseropora rosea*, the limpet *Cellana tramoserica* and the whelk *Morula marginalba* in Botany Bay, south-east Australia. Experiments were carried out using enclosures of 20 × 20 cm. Bars show SE. When *Morula* was excluded, few barnacles survived when there were either very low or very high numbers of limpets; survival was maximum at intermediate limpet densities. When *Morula* was allowed access, however, its predation pressure removed this limpet–barnacle interaction. (After Underwood *et al.*, 1983.)

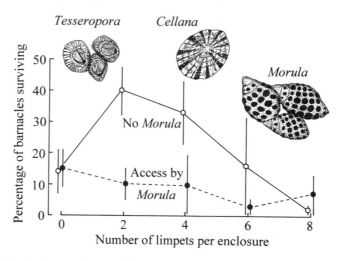

To study the inverse interaction—how barnacles affect *Cellana*—Underwood *et al.* measured the growth rates of limpets placed in fenced areas containing various densities of barnacles. Growth rates were highest with no barnacles, and declined in proportion to barnacle density: *Cellana* does not in fact normally inhabit barnacle-covered areas, and measurements of weight showed that they eventually die of starvation if confined there.

In contrast to this effect, the limpet *Patelloida* survives *better* with increasing numbers of barnacles, in the absence of whelks. Because *Patelloida* is smaller, it can feed between the barnacles, whereas the larger *Cellana* presumably has difficulty rasping with its radula on very rough barnacle-covered surfaces. As with *Cellana*, though, when *Patelloida* is enclosed together with the whelk *Morula*, the whelks eliminate any effect of barnacle density because of their high predation rate.

From these experiments, it is apparent that there are major interactions amongst all the component species in the system. Density of barnacles affects both limpet species, and in the case of *Cellana* this is neither a linear nor an all-or-none effect, but one which is beneficial at intermediate densities. *Cellana* has important effects on the survival of barnacles. The whelks affect density of both the barnacles and *Patelloida*. The outcome of these interactions evidently varies with position on the shore, exposure, season and weather, but no interactions can be classified as 'weak' in the sense of Paine, as discussed in the previous section. Underwood *et al.*'s interpretation of this system is thus quite different from a keystone-species concept. Here the various members of the community are seen to interact to varying degrees, but no overriding consumer–resource interaction is seen to dominate.

There is, of course, no reason why the communities of rocky shores in eastern Australia studied by Underwood *et al.* should necessarily be controlled in the same way as those in America, New Zealand and Chile studied by Paine. It could well be that, in places where one species tends to form monocultures, there is a tendency for keystone predators to evolve. But once again, we come to a question of scale. Could keystone species evolve where patches of monoculture are, say, centimetres or metres in diameter? Or must they be hundreds of metres or kilometres in size? Until more studies have been made of the geographical and temporal scales over which particular interactions occur, discussion will continue.

Community structure: 'top-down' and 'bottom-up' factors

Another way of thinking about the organization of communities in more general terms is to assume that they involve many interactions, some trophic and others competitive; and then to ask whether the lower trophic levels depend more upon the actions of organisms in higher levels ('top-down' factors), or whether the higher trophic levels are constrained by variation in production at lower levels ('bottom-up' factors). There seems, in fact, little doubt that both these types of factor can influence communities. For instance, some 'bottom-up' factors such as the level of primary production may determine how many trophic levels can exist, and this will affect community structure directly—although there is little evidence to show how widespread this kind of control may be. Classical 'top-down' factors might be the effect of sea urchins on kelp beds, whether or not the urchins are considered keystone species—and such factors are thought to be common.

Menge (1992) considered how these factors might act, and interact, by studying two sites on the coast of Oregon in North America. Each site has both exposed and sheltered areas. At Boiler Bay, benthic plants dominate shores at all exposures. At Strawberry Hill, plants dominate in shelter while mussels and barnacles dominate at exposed sites, where they are accompanied by many predatory starfish. The exposed shores at

Strawberry Hill have not only higher abundances of mussels and barnacles, but higher mussel recruitment, higher growth rates of mussels and barnacles, and higher predator levels. Menge hypothesized that this could be due to bottom-up factors such as higher food supply for the suspension-feeding invertebrates, possibly via higher nutrient levels and higher planktonic primary productivity. Without evidence about these factors the argument cannot be taken further. But it is apparent from Menge's study that rocky shores can no longer be studied in isolation from the processes going on in the neighbouring sea. Not only does the sea provide propagules of sessile forms, thereby governing recruitment levels, but it determines the food supply of many adult species. In the case of sessile invertebrates this is mainly via phytoplankton, but in the case of algae it may be that nutrient supplies are themselves direct controllers. Whether these bottom-up factors are as important as top-down ones is an important issue for the future.

Investigations of community structure

Communities do not usually change rapidly unless under great stress. It is, therefore, very difficult to investigate the processes that control them in short-term experiments. Below we give some options, differing greatly in the time needed for their execution.

Food webs on exposed and sheltered shores

A good introduction to trophic organization can be gained by listing the dominant or most common fauna and flora, then attempting to categorize them as producers, herbivores, predators, etc. This can be done using reference books, by examining mouthparts, and looking at gut contents. Depending upon time available, this project can examine different levels on the shore as well as shores of different exposure. Theoretical food webs can be constructed by adding in the effects of organisms known to be present for only part of the time.

Mosaic structure of exposed shores

On mussel- and barnacle-dominated shores, examine the difference in community composition between areas with mussels and those with barnacles or with bare rock patches. Clearing small quadrats will allow identification and sorting in the laboratory. If return visits can be made, mapping the extent of mussel communities and how they change provides insight into the dynamic nature of the community.

Succession

If successive visits can be made to a site, small clearance experiments can be used to demonstrate the details of succession. Clearance of grazers, for instance, may allow a succession of algae to settle. Clearance of mussels may allow settlement of other suspension feeders, such as barnacles. Areas cleared should be small, so as not to interfere with the overall habitat, and clearance of *Ascophyllum nodosum* should not be undertaken because of its long life-span.

9 Biodiversity, pollution and conservation

Chapter 8 concluded by talking about questions of scale, and whether individual species could control communities over small areas. To expand this discussion, we could consider what happens over wider areas: how do communities vary on the scale of, say, changing latitude? How do they vary globally? How does this distribution vary with time? There are several ways of approaching these broad scale problems. We begin by considering the phenomenon of variation in numbers of species from one area to another—variation in biodiversity. We go on to consider the forces such as pollution that tend to reduce this diversity. A third approach is to consider how diversity can be maintained in the future—and here we discuss the relevance of conservation.

Biodiversity of rocky shores

Although biodiversity is a major topic with regard to terrestrial ecosystems, there has been relatively little discussion of its importance in the sea, and the discussions that have occurred have mostly focused on the deep sea and the pelagic regions. Biodiversity in the sea may be defined as 'the variability among living organisms from all sources, and the ecological complexes of which they are a part; this includes diversity within species, between species and of ecosystems' (Angel, 1993). How to interpret this for rocky shores, or indeed how to measure it, is another question. Many authors have concentrated upon the number of species present, and this is one of the most important features. As we shall see later, however, the diversity of habitat types may also be very important.

Variations in biodiversity

Marine systems as a whole are exceedingly rich in higher taxa—they have representatives of 28 phyla, compared with only 11 on land. Within the oceans, however, there are great differences in species diversity between habitats. Pelagic species show relatively low diversity, while those of the benthos have been estimated to total around 10 million. Within the benthos, it has long been thought that the diversity of epifauna (animals living on the bottom but not burrowing into it) increases towards the tropics and decreases towards the poles; but this general trend, or cline, may not hold for all animal taxa (Clarke, 1992). In any case, the reasons for the cline are not understood at all, just as the reasons for a similar cline on land are unknown. In addition, there are two entirely separate phenomena underlying global patterns of species richness. First, there is the matter of the evolution and spread of new species. Secondly, there is the matter of the persistence of species in ecological units. We shall not discuss the first point, which is beyond the scope of this book, but the second point concerns many of the themes already raised, and requires further discussion.

Why should a latitudinal gradient in species diversity exist? No doubt there are some direct effects of physical factors such as climate. Ice scour will eliminate most of the invertebrates on the shore in the far north and south, and desiccation will have controlling effects in lower latitudes. Some of the changes with latitude might relate to the distribution of grazing gastropods, as well as to limitations caused by light intensity and other physical factors: the diversity of grazers is higher in the south than in the north, so that algae have less chance of escaping from grazing in southern regions. Besides this, as we have emphasized on p. 54, the activity of grazers is partly held in check by the anti-herbivore defences of the algae, and these differ with latitude. Grazers in Norway experience mostly the fucoid algae, which are defended by polyphenolics; while those in the Canary Islands meet red algae, which are defended by phenols and terpenoids. However, the factors that determine the balance between various species of algae and grazers remain elusive, and there is no overall explanation for latitudinal gradients in diversity.

Community factors affecting biodiversity

The action of grazers and predators in affecting diversity has been raised in Chapter 8 under the heading of keystone species (p. 183). Consumers that have 'strong' links with their resources may control the presence or absence of whole suites or 'modules' of other species. Thus, when sea urchins are present in the low-shore zone in California, the diversity of algae is high. When urchins are rare, the dominant alga *Chondrus crispus* outcompetes other species and forms a monoculture so that algal diversity decreases. Similarly, when the predatory starfish *Pisaster ochraceus* is present, diversity of invertebrates is high, but when the starfish is removed, diversity falls to two species because the mussel *Mytilus californianus* forms sheets which exclude other species.

However, grazers and predators do not necessarily act in this 'keystone' way. In New England, grazing activities of the snail *Littorina littorea* lead to an increase of algal diversity within tide-pools, but a decrease outside the pools (p. 83). The results outside the pools probably occur because *Littorina* consumes the green alga *Enteromorpha*, which is *not* the competitively dominant species. The grazing therefore does not have any 'keystone' effect, but simply decreases diversity because it eliminates the preferred food plant from the species list.

Further effects of grazers on algal diversity concern their 'gardening' activities (Branch *et al.*, 1992). On the west coast of North America, and in southern Africa, several species have been shown to graze algal 'gardens' around their home scar (p. 88). Within each of these gardens, algal diversity is always low: limpets usually cultivate only one algal species, which must be fast-growing. However, the diversity of the community as a whole is usually boosted by gardening species because the algae typical of the gardens are seldom found outside the gardens, and a mosaic of different algal species is created by the variety of different limpet species. Exclusion of gardening limpets on the low shore in southern Africa therefore decreases algal diversity from about 23 to 15 species (Fig. 9.1).

Fig. 9.1   Algal species richness on the shore in South Africa, in relation to limpet 'gardens'. There are many more algal species outside gardens than inside (histograms show the average for four species of limpet). When limpets are experimentally excluded, however, algal species richness declines in the low-shore zone overall, because each limpet species promotes the growth of particular algae. (After Branch *et al.*, 1992.)

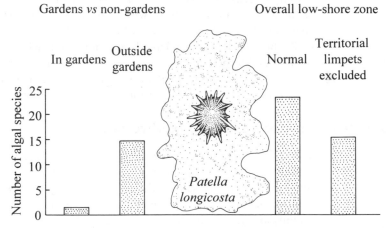

Fish may also act as grazers, and some tropical species may be termed gardeners. In contrast to limpets, they increase algal diversity within the gardens because their grazing pressure is moderate compared to that outside. This, of course, has consequences for the fauna: meiofauna of fish gardens tends to be diverse, adding to total biodiversity.

Overall, grazing species raise algal diversity when grazing pressure is relatively low, but decrease it at high grazing pressure. Are predation and grazing the major determinants of diversity on the shore? Even if they are, what are the factors that in turn dictate the distribution of the keystone predators and grazers? Why do these vary with latitude? These are some of the questions that investigators of biodiversity must tackle.

The influence of invading or 'alien' species

In some areas, major components of the ecosystem are relatively recent invaders. These 'aliens' may have very significant effects on the structure of the communities, and their biodiversity. On the east coast of North America, for instance, the grazing gastropod *Littorina littorea* is a dominant controller of algal diversity on many shores (p. 84). Yet this species was introduced from Europe in the nineteenth century. How different were American shores before its arrival? It is probable that long-lived algal species such as *Ascophyllum nodosum* were more abundant, and there may have been less 'free space' available for other colonizers. Whether this would have produced higher diversity (because of the more varied habitats available) or lower diversity (because of the takeover by monocultures of *Ascophyllum*) is debatable.

Approximately 50 species have been added to the marine fauna of Britain in the past century. Some of these have been extremely successful. The barnacle *Elminius modestus* from Australasia is now widespread on rocky shores, and a consistent member of the community at sheltered sites. Within oyster beds, several species of oysters (*Crassostrea* spp.) and one of their predators, the whelk *Urosalpinx cinerea*, have been imported from North America and are abundant. Of the algae, *Codium fragile* has become widespread, as it has done in eastern North America (p. 70). *Sargassum muticum* from Japan has probably caused most alarm amongst those wishing to conserve the present status of rocky shores (p. 58). In Normandy, *Sargassum* has been present for more than 13 years, and at its maximum development has an estimated biomass of 19 000 tonnes. On some shores it appears to have outcompeted *Laminaria saccharina*, and to have replaced some areas of the eel grass *Zostera marina*. While it has therefore added to the overall biodiversity of the coastline, it has reduced diversity on specific shores.

The extent of alien introductions in coastal areas throughout the world is probably much larger than usually accepted (Carlton, 1989). Introductions have been occurring for hundreds of years, by a variety of mechanisms: fouling or boring organisms on the surfaces of ships, together with those taken internally in ballast, and those transferred with commercial fisheries products, may number in the thousands of species. Despite international recommendations that countries should attempt to prevent such introductions, the number is apparently growing; and most species are impossible to remove once established.

Pollution and its effects upon biodiversity

The effects of human influence have already been touched upon many times in this book, but we now come to a consideration of their overall importance for coastal communities. Sudden and catastrophic occurrences such as oil spills and their treatment, or longer-term influences such as the effects of heavy metals on gastropods (p. 144) may have major impacts on the rocky shore ecosystem, and we deal with them first. There are, however, many less obvious influences, and we come to these, and how to monitor them, next. Lastly, we point to some of the ways of minimizing pollutant effects.

Before we do this, however, we need to define what is meant by pollution. Pollution of the sea can be defined broadly as 'the introduction by man of substances or energy resulting in deleterious effects'. This differentiates pollution from contamination, which is the elevation of the concentration of substances above natural levels. Evidently, the difference between the two rests upon the definition of the word deleterious, which is bound to be subjective.

Types of marine pollution vary from domestic waste with high oxygen demands and industrial effluents containing heavy metals, pesticides or radioactivity, to accidental oil spills and dumping of solids. Related problems may arise from reclamation, but this is much more common on sedimentary shores than on rock. We go on to discuss some examples.

Gross accidental pollutant effects

The wreck of the oil tanker *Torrey Canyon* in 1967, and subsequent effects of the oil and toxic dispersants, have already been described briefly (p. 82). The oil itself was not very toxic, and most damage was done by the spraying of detergents containing aromatic solvents and surfactants. These killed practically all the fauna and damaged the macrophytic flora, although in some places some of the gastropods and some of the fucoids survived. Immediately after the clean-up operations, species diversity declined dramatically, often to zero. Recolonization showed the usual sequence following gross disturbance (p. 164), and species numbers gradually built up again. Initially, while species numbers were low, the total standing crop or biomass was very high, due to growth of algae. This declined again as numbers of grazing animals increased. The time scale involved in recovery was variable, depending upon the degree of initial damage. Some of the least affected areas recovered in 5–8 years, while the sites that received repeated applications of dispersants were not back to normal after 10 years (Southward and Southward, 1978).

Sudden incidents of gross pollution thus act as disturbance mechanisms: successions are broken and must start again with early colonists. Species diversity is inevitably reduced, and it may take many years for the community to recover. A generalized diagram suggesting the time-scale of changes that may occur is given in Fig. 9.2.

The contrast between the *Torrey Canyon* spill, and those in which dispersants were not used, is extreme. In 1972, the troopship *General*

M.C. Meigs was wrecked on the coast of Washington. The initial major spill of fuel oil was followed by 5 years of continual low-level release of oil. There was little immediate major damage to much of the intertidal fauna, though abundance of barnacles, mussels and colonial anemones declined steadily until 1973. The sea urchin *Strongylocentrotus purpuratus* was severely damaged—up to 70% of the population lost their spines, and many died—but the population subsequently recovered rapidly. Of the flora, *Laminaria andersonii* lost fronds and showed damage even 2 years later, while other species were bleached. By 1977, though, most algal growth appeared normal again, and it was concluded that no major change in community balance had occurred.

Fig. 9.2   Generalized graph showing effects of a major oil spill on abundance and diversity of species. Opportunist species may bloom very rapidly, but it may take many years for rare and sensitive species to return. (Partly after Suchanek, 1993.)

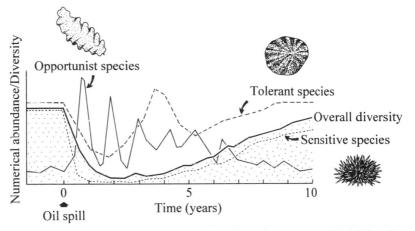

There was even less response to oil when the tanker *World Prodigy* released 922 tonnes of fuel oil on the coast of New England in 1989. Growth rates of *Laminaria saccharina* and *L. digitata* were unchanged, apparently because little fuel oil was mixed into the water column. However, sublethal effects of oil have been well documented for a variety of invertebrates (Suchanek, 1993). Cnidarians show abnormalities in their morphology, molluscs show reduced feeding rates and loss of energy for growth and reproduction, while some crustaceans produce fewer larvae when held in oiled sediments than do controls.

One of the problems in assessing the damage caused by relatively low-level pollution is that, as we have emphasized (p. 161), communities may show wide variations from purely natural causes. This means that detecting change is a far from simple matter, and we take up the theme of monitoring in the next section.

Chronic pollution and problems of detection and monitoring

The majority of pollutants in the sea are probably released via effluent pipes and into rivers rather than in catastrophic accidental spills. The effects of this continuous 'chronic' pollution are very hard to assess, in the face of normal background changes. There are various methods of analysing changes in community structure, whether these are brought about by natural effects or by pollution. They fall into three categories, each of which has its advantages (Warwick and Clarke, 1991). First, 'univariate' methods relate abundances of different species to a single index such as the Shannon–Wiener diversity index. This index takes into account both species numbers and abundance. It rises with increase in species numbers in the ecosystem, and so gives an easily comprehensible measure of diversity. Secondly, graphical methods plot relative abundances as curves. Typical of these are 'k-dominance' curves, which plot cumulative percentage of overall abundance, arranged against the rank order of species (arranged in order of decreasing abundance). Raising of the curves shows that fewer species contribute more to the community, and suggests a polluted situation. The third type of method is ' multivariate' analysis. This compares the communities on the basis of individual component species they have in common as well as similarity in terms of overall species abundance. Multivariate analyses such as multi-dimensional scaling ordination (MDS) are very sensitive and good at discriminating between different sites or times, but are difficult to interpret in terms of possible causes.

Whichever method is used, it is important to minimize the influence on a survey of 'normal' background changes. One way of doing this is to choose a specific microhabitat which one would expect to contain similar communities everywhere. Kelp holdfasts, in particular, have proved useful in monitoring studies. Holdfasts contain a diverse fauna and are easily sampled. On the north-east coast of England, where sites vary from unpolluted to those polluted by domestic or industrial waste such as coal waste, heavy metals and organic loads with faecal bacteria, the fauna in holdfasts of *Laminaria hyperborea* varies enormously. On the unpolluted coast, holdfasts contain up to 91 species of animals. At polluted stations, this number is reduced to 66. Such a reduction in species to a 'pollution-tolerant' set is now recognized as typical of polluted sites. Many of the 'pollution-intolerant' species, found only at unpolluted sites, are in fact rare even where they occur; and the loss of rare species is now also recognized as a standard response to pollution.

The fauna of kelp holdfasts (*Ecklonia radiata*) was used to monitor the effects of a domestic effluent on the coast of New South Wales (Smith and Simpson, 1992). The effluent had received secondary treatment and chlorination. Using MDS, Smith and Simpson were able to show that the holdfast communities changed from dominance by suspension-feeding bivalves and polychaetes, near the outfall, to dominance by detritus-feeding amphipods further away.

This type of study does, however, raise the problem of whether the effects that are studied can strictly be called 'pollution'. Reduction in species numbers would probably be universally agreed to be a 'deleterious' effect. But a change in community structure from detritus-feeding to suspension-feeding would certainly be debatable in terms of 'desirability'. A similar example is the more general input of nutrients, which may lead to increased growth of algae, or eutrophication. Mild eutrophication may change species composition but may also increase such factors as fisheries yield. Is this to be termed pollution or fertilization? It is here that the subject of pollution ceases to be a matter of science, and becomes one of politics.

Studies on untreated sewage outfalls have shown more drastic changes. A small sewage plant on San Clemente Island, California, discharged 95 000 l of effluent per day. At the outfall, macrophytes increased on the upper shore due to heavy growth of the green alga *Ulva californica* and red alga *Gelidium pusillum*. On the lower shore, macrophytes decreased: the kelp *Egregia laevigata* was absent, and was replaced by colonies of a suspension-feeding mollusc, the vermetid *Serpulorbis squamigerus* (Fig. 9.3). Many other species were absent in the pollutant plume, so that the Shannon diversity index was reduced from 3.9 in control areas to 2.8 at the outfall.

Fig. 9.3   Distribution of the dominant fauna and flora on the shore at San Clemente, California, in relation to an outfall of untreated sewage. Areas of the opportunist alga *Ulva* sp. and of the vermetid mollusc *Serpulorbis squamigerus* dominate the mid shore near the outfall, replacing the algae *Corallina* sp. and *Egregia laevigata*. The low-shore zone of the alga *Eisenia arborea* is relatively unaffected. (After information in Littler and Murray, 1975.)

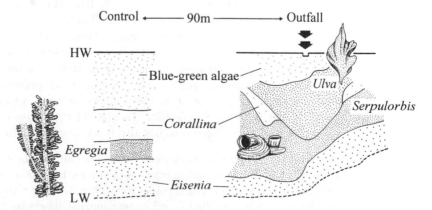

In Argentina, where a much larger outfall produces about 14 million l per day, the control areas are dominated by a mussel, *Brachidontes rodriguezi*, and by *Ulva lactuca* in tide-pools. At the outfall, there is much more bare space, more blue-green algae, *less Ulva*, and no *Brachidontes*. Species diversity is reduced at the outfall, giving a Shannon index of 1.6, compared with 2.2 at the furthest distance monitored. However, at intermediate distances, diversity rises to 2.7, suggesting that a low level of pollution actually raises diversity (Gappa *et al.*, 1990). The explanation for this may lie in the fact that *Brachidontes*, like many other mussels, is competitively dominant and forms monocultures. Any disturbance that interferes with this monoculture acts to restart successions, and allows a mosaic of other species to appear. The action is thus similar to that shown experimentally in California, where diversity is raised by allowing the entry of starfish, the predator of mussels.

Responses to pollution

Assuming that pollution can be detected, and assuming that agreement can be reached about its undesirability, what can be done about it? The major response has always been, and still is, to base action on the old phrase 'the solution to pollution is dilution'. So planning consents are usually given based on calculated dispersal of effluent into a large body of water, preferably with tides or currents taking the effluent offshore. In theory, this procedure can cope with shore-based effluents, although in practice the money available for sewage treatment plants or long outfalls may not be available. Again, these are political, not scientific decisions.

The case of accidents, such as spills of oil or toxic chemicals, is more complex and involves contingency planning. The approaches to this planning, and their effectiveness, differ immensely between various countries (Gubbay, 1989). As an example, we take plans developed to counteract oil spills around the coast of Britain. These cover the entirety of the coastline, but have been particularly developed in the region of ports that import or export oil. Milford Haven, in south-west Wales, for instance, has developed an anti-oil-pollution 'plan', operated by a voluntary organization which consists of the Port Authority and the oil companies. The plan ensures that when pollution occurs, action is taken immediately, and the costs of any necessary clean-up operations are sorted out later. A prime consideration is to prevent oil from reaching the shore, since it is there that the most adverse effects occur. Speed is therefore essential. In general the removal of spilled oil is preferred over dispersion, but this is not always feasible.

The plan has, in general, been very successful, but inevitably accidents have occurred in which oil has reached the shore. For example, a spill of 100 tonnes of crude oil which coincided with a flood tide and a force 8 gale from the west resulted in oil contaminating shores in most of the upper part of Milford Haven. Under these circumstances, dispersion was impossible, and use of booms would have been impracticable because of the weather. Major complements to the plan have therefore been the preparation of 'sensitivity maps' and oil-spill guidelines for the shoreline (Fig. 9.4). Such maps allow the treatment of oil spills to be carried out with sympathy to wildlife.

Fig. 9.4   Generalized 'sensitivity map' of a rocky coast, indicating habitat types and areas important for wildlife. Months for which birds are most vulnerable are shown by the filled segments in the rings surrounding the bird symbols. (After maps produced by the Nature Conservancy Council, Britain.)

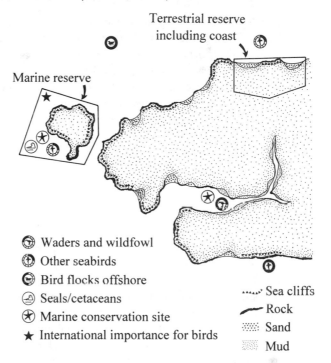

To provide more detail for the personnel actually involved in cleaning the shore, some authorities have produced 'data books' listing all accessible beaches and laying down clear guidelines for action to be taken (or not) in the case of any polluting incident. For example, these might specify whether dispersants should be used, whether beach material should be removed or not, and whether seaweeds should be torn up or cut to leave the stipes for regrowth. It is to be noted, however, that on the majority of inaccessible shores, the only (and perhaps the best) policy is to leave oil to be dispersed and weathered naturally.

The oil terminal at Sullom Voe, in Shetland, began operating in 1978, and during the 1980s handled over 1 million barrels of oil daily from the North Sea platforms. Several organizations have been involved in co-ordinating responses to possible pollution, the major one being the Shetland Oil Terminal Environmental Advisory Group (SOTEAG). This group comprises the Shetland Council and the oil companies, and is chaired by independent (university) scientists (Kingham, 1991). It operates a monitoring programme to provide early warning of any deterioration in the environment of Sullom Voe due to the operation of the terminal. This involves chemical, macrobenthic and sea-bird monitoring, and qualitative surveillance of rocky shores. As with the monitoring programmes discussed earlier, of course, these will suffer from the usual problems of attempting to detect 'unnatural' changes in environments where 'natural' changes can be immense.

As in Milford Haven, plans for what to do in the case of an emergency in Shetland are not always effective in practice because bad weather can prevent appropriate action being taken. In January 1993, the tanker *Braer* was wrecked, causing a spill of 85 000 tonnes of light crude oil. Strong winds prevented any successful rescue measures, and the oil dispersed causing extensive bird casualties. Because of the very rapid dispersion, it has been difficult to estimate overall biological damage, but the accident has caused widespread discussion of the rules that govern routes taken by tankers, and suggestions have been made that no tankers should be allowed near to shorelines except when entering and leaving oil terminals.

Conservation

The biological influences of pollution may generally be summarized by saying that pollution reduces biodiversity—even though, as we have seen above, some forms of pollution may, at mild levels, actually increase diversity. One way of attempting to maintain a high biodiversity is to undertake active conservation. In conservation areas the status of the natural ecosystem is then given a higher priority than other (human interest) factors. But what does this mean for rocky shores? How should areas be selected? What legal protection can and should be given to such sites? How should they be 'managed', if indeed this is possible, so that they will persist in the future?

Approaches to marine conservation

Marine nature conservation has lagged far behind the programme of conservation on land. Compared with the multitude of terrestrial nature reserves and protected areas (mostly set up to protect vertebrates and angiosperms), the protection of the intertidal and subtidal regions (dominated by invertebrates and algae) has so far been minimal. As a result, coastal destruction and harvesting may have led to scores of extinct populations of invertebrates, as well as the extinction of some species.

Exploitation of rocky-shore biota is widespread throughout the world. Chile and South Africa provide good examples (Siegfried, 1994). In Chile, 182 400 tonnes of algae (mainly kelps) are collected each year, and in South

Africa 30 000 tonnes are taken. The lobster catch in Chile is 500 tonnes per year, while it is 3000 in South Africa. In Chile, a muricid whelk, *Concholepas concholepas*, was harvested to a maximum of 11 000 tonnes per year, but this resulted in such a population decline that closed seasons were introduced, and then total bans on collection. In South Africa, overfishing of the lobster *Jasus lalandii* necessitated introduction of catch quotas, and a minimum size of the individuals caught, but even so the populations have been severely reduced, and now the catch quotas are not being reached. Changes in the size-frequency composition of exploited populations are well shown in the abalone of South Africa (Fig. 9.5).

Fig. 9.5 Size composition of the abalone, *Haliotis midae*, at two sites in South Africa. Very few individuals of catchable size are still present at the exploited site. (After Dye *et al.*, 1994.)

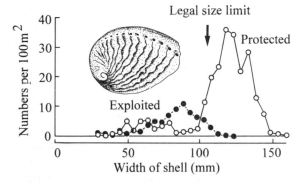

Several management strategies can be considered for exploited populations. One option is to rotate the areas open for exploitation so that populations can recover in the intervals between cropping. Another is to create permanent sanctuaries for individual species. If exploitation is allowed, it is possible to control the effort devoted to the catch in some way—by the use of quotas, bag limits, mesh sizes or licences. Reproduction of the population can be encouraged by imposing closed seasons or bans on the collection of reproductive stages, and by harvesting only above a minimum size. Lastly, there is the exciting possibility that mariculture can be used to reseed denuded areas. This procedure is being tried in Chile for the whelk *Concholepas* referred to above.

'Coastal zone management', as it is frequently called, has now emerged into the political scene in North America and in Europe, and there is widespread perception that marine flora and fauna require protection. The essential problems are thus to decide first 'what needs conserving?', and secondly 'how can this best be done?'

There is probably little antagonism to the idea that biodiversity should be conserved. This means maintaining a wide diversity of habitat types, since most marine biota are governed in their distribution by the availability of appropriate physical niches. It should be emphasized that this involves not just conservation of areas rich in species, but also of some areas that themselves have only low diversity. Exposed rocky shores come into this category, since they have a low species diversity, but support several species that are restricted to areas of high wave exposure. The communities there, although not diverse, are unique. Conservation also involves sustainability—so that habitats and the species within them are self-perpetuating.

The general principles that should be applied to bring about such a conservation strategy can be considered in three tiers (English Nature, 1993). To ensure that the overall environment of the sea remains suitable for marine organisms, the first tier of conservation is to look after a wide area of sea. Here the objectives are to maintain or improve water quality, to minimize pollution and to regulate fishing, recreation and development. The second tier is to look after wide areas of coastline. This can best be done by raising awareness about the importance of marine wildlife and developing programmes of co-operative management. The third tier is to focus on areas of special interest and this must involve providing statutory protection for them.

The tiered approach has its parallels in the UNESCO programme 'Man and the Biosphere'. Under this programme, large 'biosphere reserves' have been set up world-wide, each, in theory, having a series of zones (Clark, 1991): a core area, in which natural ecosystems are strictly conserved; a surrounding area which can be used for 'non-destructive activities'; and a third, outer ring, the transition zone, which has a flexible boundary, allows some 'economic use' and has fewer controls (Fig. 9.6).

Fig. 9.6   Generalized map of zoning in a 'biosphere reserve'. (After descriptions by Clark, 1991.)

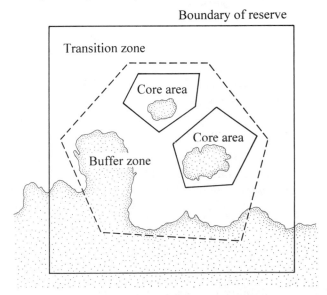

How does one decide which marine areas should be conserved? To determine suitable areas, large-scale survey work is required, in which sites appropriate for conservation are determined. Three main criteria can be used: first, the diversity of habitats, communities or species; secondly, a consideration of whether the sites are good examples of communities; and thirdly, whether they are rare or unusual. Two phases of survey are usually required: an initial phase in which a broad overall view of the coastline is gained; and a second phase in which detailed biological and physical data are gathered and then analysed to define the types of communities and their distribution. Such a process can take many years.

Following the proposal of criteria by the International Union for the Conservation of Nature (IUCN), there has been remarkable overall agreement about the priorities to be adopted after initial surveys have been carried out (Gubbay, 1988). The IUCN suggests eight categories that should be considered. Ecological and research/education criteria must be considered together with social, economic, landscape, cultural, regional and practical matters. Thus diverse aspects such as fishing rights, threats to the site and its fragility must be integrated into the approach. Here we discuss only those criteria that relate to ecology.

One criterion that is common to most programmes is diversity. For example, the island of Skomer, in Wales, has a diversity of environments, and because it has species from both northern and southern regions, it is particularly species rich, and has been made a marine reserve. Two other factors indicate the ends of the conservation spectrum: representativeness and uniqueness. In general, sites with communities that are particularly good examples of their type, either internationally or more locally, are agreed to be worth conserving. For instance, the area around St. Abbs Head in north-east England is representative of classic kelp forest and grazed *Echinus* areas, and is a voluntary marine conservation area. On the other hand, the maerl beds of Britain (p. 73) are common only in north and west Scotland, but the largest bed in Britain is in south Cornwall, and this unique area has also been protected by inclusion within a voluntary marine conservation area.

The presence of rare or endangered species is another important factor. On rocky shores many species that are apparently 'rare' have large sublittoral populations. In more isolated habitats such as coastal lagoons, however, species often exist in very small populations. In Britain, the bryozoan *Victorella pavida* is restricted to one lagoon in Cornwall, while the anemone *Edwardsia ivelli* is found only in one lagoon in Sussex.

A final major question for discussion is the size and frequency needed for marine reserves. Conservation in the sea needs a large-scale policy because, as emphasized above, the boundaries of any reserve are highly permeable, and conditions in a large area of sea may be critical. For organisms with pelagic larvae, it has been suggested that dispersal distances may be up to 10^7 m, indicating that enormous areas need protection. On the other hand, small-scale experiments have suggested that relatively small refugia may prevent species extinction by providing a continuing source of breeding adults (Quinn *et al.*, 1993). On this basis, smaller, closely spaced reserves, each within the dispersal range of the larvae, might be more appropriate than few large areas. The topic will doubtless provide future discussion under the title SLOSS—'single large or several small?' One model for a population of fish, *Coracinus capensis*, on the southern coast of South Africa, suggests that export of individuals would be greatest from reserves between 20 and 60 km in length. Three separate reserves of, say, 50 km in length would therefore be of greater benefit to anglers than a single reserve of 150 km. Such modelling exercises may provide at least some quantitative basis for determining the extent and arrangement of reserves.

Examples of marine nature reserves

Lough Hyne, in Ireland, was declared Europe's first statutory marine reserve in 1981. It is a sheltered sea lough with a predominantly rocky shoreline, which has attracted research workers since the late nineteenth century, but particularly from the 1920s onwards. The biology of the lough has therefore been investigated in detail, and one of the major aims of the Irish Wildlife Service, which manages the reserve, is to promote further understanding of the lough's ecosystems. Because the lough is small (roughly 1 km^2), it is vulnerable to damage, and research is allowed only by permit. Before it was a statutory reserve, damage caused by research was limited by observance of a simple rule: the shore of the lough was divided into sectors, and no activity was allowed if it changed more than 10% of any particular sector.

Despite its small size, the Lough Hyne Marine Reserve has an extraordinary diversity of habitats. Within the lough itself the sheltered shoreline varies from vertical bedrock to scattered boulder beds, with patches of gravel and saltmarsh. In the channel to the sea, there is a section with fast-flowing rapids (reversing at every tide) and a gradient of increasing exposure to wave action. On the outer coast, exposure is extreme, but parts of the shore support a complex system of large tidepools. The high diversity of fauna and flora is partly determined by this diversity of habitats, but partly also by the lough's geographical position between the cold-temperate (Boreal) zone to the north and the warm-temperate (Lusitanean) zone to the south. Some of the species in the lough are in fact more typical of the Mediterranean region than the Atlantic coast (Lusitanean) region, and these are probably relicts from warm interglacial times.

The major intent of the management plan for Lough Hyne is to 'ensure the conservation of the marine ecosystem' (O'Donnell, 1991). Research is subordinate to this function, as are the needs for education, recreation and commercial utilization. Thus, power-boating is not allowed, while swimming and angling are allowed. Scuba diving is regulated by a permit system. Commercial fishing is allowed only for local families who have traditional fishing rights, and mariculture is not permitted.

In 1986 the island of Lundy became a statutory reserve, the first in Britain. Like Lough Hyne, Lundy's shores offer a wide diversity of habitats, and also harbour a mixture of Boreal and Lusitanean species. Unlike Lough Hyne, the intertidal zone is wide, well suited to study, and shows classic patterns of zonation. The Lundy reserve is also much bigger than Lough Hyne. The island is about 5×1.3 km, and the total reserve is more than 4×6 km. The proposed scheme for the reserve is an excellent example of zoning. There are five zones altogether (Fig. 9.7). An outer 'general use' zone allows commercial trawling and dredging. An inner 'refuge' zone prohibits these but allows the use of fixed-net fisheries and potting for lobsters and crabs. A 'sanctuary' zone permits no commercial fishing except limited potting, and a 'recreational' zone is similar but permits snorkelling and swimming. Archaeology 'exclusion' zones allow no activities except research, and that only by a permit system. Diving is otherwise allowed over the whole reserve, but spear-fishing is prohibited.

Fig. 9.7   Zoning scheme for the Lundy Island Marine Nature Reserve, England. (After maps produced by English Nature.)

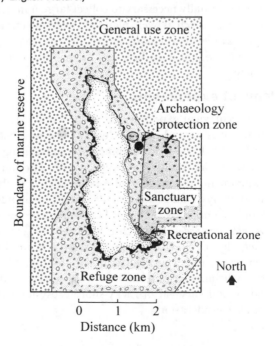

Some practical guidelines for conservation when working on rocky shores

Many rocky-shore communities, despite their robust tolerance of physical forces, are fragile when faced with human interference. Below we mention some of the problems caused by working on the shore, and give some brief guidelines for minimizing deleterious effects.

The importance of replacing boulders

Some of the richest communities are found where stable boulders provide a variety of microhabitats. Often these take the form of algal-dominated communities on upper surfaces and encrusting faunal communities, together with more mobile animals, below. It is essential that if such boulders are disturbed, they are replaced carefully in their correct orientation. Otherwise both sets of communities will be obliterated, and it may take up to 10 years for them to recover, as shown by many of the experiments in this book that have recorded the effects of major disturbance.

The problems of trampling

Trampling by large parties, or by repeated visits of small groups, may destroy or alter communities just as effectively as the overturning of boulders. Especially in monitoring exercises, or demonstrations to large classes, these effects should be assessed and minimized.

The problems of overcollecting

It is not usually necessary to collect large numbers of specimens from the field if research is well planned; overcollecting has certainly reduced species diversity on many shores. Because some species reproduce slowly, careful thought must be given to how many specimens can be removed safely.

The problems of experiments

Any ecological experiment must, to some extent, alter the ecosystem in which it is carried out. Exclusion cages or fences, removal experiments and the addition of equipment for measuring physical and chemical variables, all modify the environment. The numbers and areas of such experiments should therefore be restricted to those necessary for statistical analysis, and experiments should be avoided if long-term damage would be inflicted by them. For example, some species, such as the alga *Ascophyllum nodosum*, which is harvested for extraction of alginates, have lifecycles of 20 years or more. Recolonization by this alga may be only very slow, and experiments upon it should not be carried out unless some overwhelming reason can be put forward to justify them. In general, the safety and survival of rocky-shore ecosystems should be placed before other considerations.

Appendix I A brief classification of selected organisms

ALGAE
  PHAEOPHYTA (brown algae)
   Laminariales (kelps)   *Laminaria, Ecklonia,*
                              Macrocystis

   Fucales (fucoids)   *Fucus, Durvillaea*

  CHLOROPHYTA (green algae)
   Ulvales   *Ulva, Enteromorpha*

   Codiales   *Codium*

  RHODOPHYTA (red algae)
   Cryptonemiales   *Corallina, Lithophyllum*

   Gigartinales   *Gigartina, Mastocarpus,*
                           Chondrus

BACILLARIOPHYTA (diatoms)

Achnanthes, Navicula

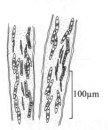

100µm

PORIFERA (sponges)

Scypha (= *Grantia*)

CNIDARIA (coelenterates)
  HYDROZOA (hydroids)

Nemertesia, Dynamena

ANTHOZOA (sea anemones)

Actinia, Anthopleura

ANNELIDA (segmented worms)
  POLYCHAETA

*Pomatoceros, Sabellaria,
Spirorbis*

MOLLUSCA (molluscs)
  POLYPLACOPHORA (chitons)

Katharina

GASTROPODA (gastropods)
Fissurellidae (keyhole limpets)

Fissurella

Patellidae (patellid limpets)

Patella, Cellana

Acmaeidae (acmaeid limpets)   *Acmaea, Lottia*

Siphonariidae (pulmonate limpets)   *Siphonaria*

Neritidae (nerites)   *Nerita*

Trochidae (topshells)   *Gibbula, Tegula, Monodonta*

Vermetidae   *Serpulorbis*

Muricidae (whelks)   *Nucella, Thais, Morula, Concholepas*

Nudibranchia (nudibranchs)   *Doridella, Aeolidia, Archidoris*

BIVALVIA (bivalves)   *Mytilus, Ostrea, Brachidontes*

CRUSTACEA (crustaceans)
CIRRIPEDIA (barnacles)   *Semibalanus, Chthamalus,*

Tesseropora,

Pollicipes

ISOPODA (isopods)   *Ligia, Ligyda, Idotea*

AMPHIPODA (amphipods)   *Hyale, Caprella*

DECAPODA (crabs and lobsters)   *Carcinus, Cancer,*
Cyclograpsus, Homarus

BRYOZOA (bryozoans or sea-mats)   *Electra, Alcyonidium,*
Membranipora

ECHINODERMATA (echinoderms)
ASTEROIDEA (starfish)   *Pisaster, Asterias*

ECHINOIDEA (sea urchins)   *Paracentrotus, Echinus*
Strongylocentrotus

CHORDATA (chordates)
  UROCHORDATA (tunicates)
  Ascidiacea (sea squirts)   *Pyura*

VERTEBRATA (vertebrates)
Teleostei (bony fish)   *Lipophrys* (shanny)

Aves (birds)   *Arenaria* (turnstone)

Appendix II   Some sites at which research quoted in the text has been carried out

Figure A shows sites on a world-wide scale; Fig. B shows sites in north-west Europe.

A

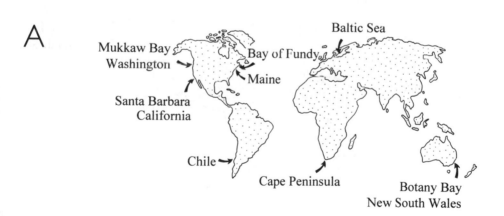

B

Sullom Voe

Norway

Isle of Cumbrae

Denmark

Isle of Man

Menai Straits

Helgoland

Baltic Sea

Lough Hyne

Lundy

Milford Haven

Severn Estuary

Plymouth

Roscoff

La Rance

Further reading

General

A wide background to marine biology is provided by Barnes and Hughes (1988), who place the study of rocky shores into context. Lewis (1964) gives a classic account of rocky shores, based on Britain but useful for the whole of north-west Europe. In a multi-author volume, Moore and Seed (1985) provide a variety of recent views. Hawkins and Jones (1992) give an introductory guide to field courses, with sections on identification and class exercises. Newell (1979) provides a compendium of information about the biology and physiology of littoral fauna.

For methods, refer to Baker and Wolff (1987), who give some detailed accounts and references to others.

Tides and waves

Pethick (1984) discusses the causes and properties of tides and waves. Denny (1988) covers in detail the problems for organisms of waves on the shore.

Identification of fauna and flora

Classic regional accounts of shores and their inhabitants are often useful for identification. For the Pacific coast of North America, see Ricketts and Calvin (1952). For the Atlantic coast of North America, see Gosner (1979). For New Zealand see Morton and Miller (1973). For Australia see Dakin (1980). For north-west Europe, see Yonge (1949) and Barrett and Yonge (1968).

In Europe, there is a great variety of up-to-date guides that can be used to identify most organisms to species level with reasonable certainty. These vary from small volumes designed for use on the shore (Quigley and Crump, 1986) to more detailed guides such as Campbell (1994), Fish and Fish (1989), Hayward (1995), and Hayward and Nelson-Smith (1996), and expensive reference volumes strictly for use in the laboratory (Hayward and Ryland, 1990).

For more complete coverage and further detail, there are several series of specialist guides useful in north-west Europe. *Faune de France*, published by the Librairie de la Faculté des Sciences, Paris, is old but particularly useful for polychaetes and bryozoans. The *Synopses of the British fauna*, produced by the Linnean Society of London and the Estuarine and Coastal Sciences Association, aim eventually to cover all marine species. Information about these can be obtained from Field Studies Council Publications, Preston Montford, Shrewsbury SY4 1HW, UK. Another set of guides is produced by the Field Studies Council, mostly under the title AIDGAP (aids to identification of difficult groups of animals and plants). These cover the red and brown seaweeds well. Information about the AIDGAP series may also be obtained from the Field Studies Council, which produces a variety of other guides and offprints. The most detailed guides to seaweeds are produced in collaboration by the British Phycological Society and the British Museum. Information about these may be obtained from the British Museum (Natural History), Cromwell Road, London SW7 5BD, UK.

Shore communities

Communities around the world are described by Stephenson and Stephenson (1972), and a very detailed account is given for Britain by Lewis (1964).

Biology of algae

General accounts of the algae are given by Dring (1982) and by Bold and Wynne (1985). Reproduction is reviewed by Santelices (1990*b*), and general ecology by Chapman (1986).

Biology of grazers

Grazing of algae by marine invertebrates is reviewed by Hawkins and Hartnoll (1983). Branch (1981) reviews the biology of limpets. The volume edited by John *et al.* (1992) contains a variety of papers dealing with benthic plant–animal interactions.

Biology of suspension feeders

The problems of fouling are reviewed in a multi-author book edited by Costlow and Tipper (1984). The mechanics of water flow and how these relate to factors such as feeding in suspension feeders are discussed by Denny (1988). Jørgensen (1966) describes the physical variety of suspension-feeding mechanisms, but note that many interpretations of function have changed. The biology of barnacles is reviewed by Anderson (1993) and biology of sea anemones by Shick (1991).

Biology of predators

The biology of dog whelks is reviewed by Crothers (1985) and that of the crab *Carcinus* by Crothers (1967, 1968). For a discussion of starfish and their food, see Sloan (1980). Nudibranchs and their biology are covered by Thompson and Brown (1984). The influence of birds as predators on rocky shores is discussed by Feare and Summers (1985).

The functioning of communities

Competition on the shore is reviewed by Branch (1984). The ways that predators affect competition are discussed by Buss (1986). Lubchenco and Gaines (1981) discuss the effects of herbivores. For a brief discussion of supply-side ecology, see Underwood and Fairweather (1989). For an account of how experiments should be laid out, see Underwood (1991). The concept of keystone species is discussed by Foster and Schiel (1988).

Pollution and conservation

The whole subject of pollution in the sea is covered by Clark (1992). For comments on conservation and conservation sites, see Gubbay (1988).

References

Aleem, A.A. (1950). Distribution and ecology of British marine littoral diatoms. *Journal of Ecology*, **38**, 75–106.

Anderson, D. (1993). *Barnacles*. Chapman & Hall, London.

Angel, M.V. (1993). Biodiversity of the pelagic ocean. *Conservation Biology*, **7**, 760–772.

Baker, J.M. and Wolff, W.J. (eds) (1987). *Biological surveys of estuaries and coasts*. Cambridge University Press, Cambridge.

Baker, S.M. (1909). On the causes of the zoning of brown seaweeds on the seashore. *New Phytologist*, **8**, 196–202.

Ballantine, W.J. (1961). A biologically-defined exposure scale for the comparative description of rocky shores. *Field Studies*, **1**, 1–19.

Barnes, R.S.K. (1984). *Estuarine biology* (2nd edn). Arnold, London.

Barnes, R.S.K. and Hughes, R.N. (1988). *An introduction to marine ecology*. Blackwell, Oxford.

Barrett, J.H. and Yonge, C.M. (1968). *Collins pocket guide to the seashore*. Collins, London.

Bassindale, R. (1943). Studies on the biology of the Bristol Channel XI. The physical environment and intertidal fauna of the southern shores of the Bristol Channel and Severn Estuary. *Journal of Ecology*, **31**, 1–29.

Bassindale, R., Ebling, F.J., Kitching, J.A. and Purchon, R.D. (1948). The ecology of the Lough Ine rapids with special reference to water currents I. Introduction and hydrography. *Journal of Ecology*, **36**, 305–322.

Bayne, B.L. (ed.) (1976). *Marine mussels: their ecology and physiology*. Cambridge University Press, Cambridge.

Bayne, B.L. and Hawkins, A.J.S. (1992). Ecological and physiological aspects of herbivory in benthic suspension-feeding molluscs. In *Plant–animal interactions in the marine benthos* (ed. D.M. John, S.J. Hawkins and J.H. Price), pp. 265–288. Clarendon Press, Oxford.

Bell, E.C. and Denny, M.W. (1994). Quantifying 'wave exposure': a simple device for recording maximum velocity and results of its use at several field sites. *Journal of Experimental Marine Biology and Ecology*, **181**, 9–29.

Bertness, M.D. (1984). Habitat and community modification by an introduced herbivorous snail. *Ecology*, **65**, 370–381.

Bertness, M.D. (1989). Intraspecific competition and facilitation in a northern acorn barnacle population. *Ecology*, **70**, 257–268.

Bold, H.C. and Wynne, M.J. (1985). *Introduction to the algae. Structure and reproduction* (2nd edn). Prentice Hall, New Jersey.

Brace, R.C. and Reynolds, H.A. (1989). Relative intraspecific aggressiveness of pedal disc colour phenotypes of the beadlet anemone, *Actinia equina*. *Journal of the Marine Biological Association of the UK*, **69**, 273–278.

Branch, G.M. (1981). The biology of limpets: physical factors, energy and ecological interactions. *Oceanography and Marine Biology Annual Review*, **19**, 235–380.

Branch, G.M. (1984). Competition between marine organisms: ecological and evolutionary implications. *Oceanography and Marine Biology Annual Review*, **22**, 429–593.

Branch, G.M., Harris, J.M., Parkins, C., Bustamante, R.H. and Eekhout, S. (1992). Algal 'gardening' by grazers: a comparison of the ecological effects of territorial fish and limpets. In *Plant–animal interactions in the marine benthos* (ed. D.M. John, S.J. Hawkins and J.H. Price), pp. 405–423. Clarendon Press, Oxford.

Brawley, S.H. (1992). Mesoherbivores. In *Plant–animal interactions in the marine benthos* (ed. D.M. John, S.J. Hawkins and J.H. Price), pp. 235–263. Clarendon Press, Oxford.

Breen, P.A. and Mann, K.H. (1976). Changing lobster abundance and the destruction of kelp beds by sea urchins. *Marine Biology*, **34**, 137–142.

Burrows, M.T. and Hughes, R.N. (1991). Optimal foraging decisions by dog whelks, *Nucella lapillus* (L.): influences of mortality risk and rate-constrained digestion. *Functional Ecology*, **5**, 461–475.

Buss, L.W. (1986). Competition and community organization on hard surfaces in the sea. In *Community ecology* (ed. J. Diamond and T.J. Case), pp. 517–536. Harper and Row, New York.

Cambridge, P.G. and Kitching, J.A. (1982). Shell shape in living and fossil (Norwich Crag) *Nucella lapillus* (L.) in relation to habitat. *Journal of Conchology*, **31**, 31–38.

Campbell, A.C. (1994). *Seashores and shallow seas of Britain and Europe*. Hamlyn, London. (reissue of a book published in 1976.)

Carlton, J.T. (1989). Man's role in changing the face of the ocean: biological invasions and implications for conservation of near-shore environments. *Conservation Biology*, **3**, 265–273.

Castenholz, R.W. (1961). The effect of grazing on marine littoral diatom populations. *Ecology*, **42**, 783–794.

Chapman, A.R.O. (1986). Population and community ecology of seaweeds. *Advances in Marine Biology*, **23**, 1–161.

Chapman, A.R.O. and Johnson, C.R. (1990). Disturbance and organization of macroalgal assemblages in the north west Atlantic. *Hydrobiologia*, **192**, 77–121.

Chapman, M.G. and Underwood, A.J. (1992). Foraging behaviour of marine benthic grazers. In *Plant–animal interactions in the marine benthos* (ed. D.M. John, S.J. Hawkins and J.H. Price), pp. 289–317. Clarendon Press, Oxford.

Chelazzi, G., Santini, G., Parpagnoli, D. and Della Santina, P. (1994). Coupling motographic and sonographic recording to assess foraging behaviour of *Patella vulgata*. *Journal of Molluscan Studies*, **60**, 123–128.

Christie, A.O. and Evans, L.V. (1962). Periodicity in the liberation of gametes and zoospores of *Enteromorpha intestinalis* Link. *Nature, London*, **193**, 193–194.

Clark, J.R. (1991). Management of coastal barrier biosphere reserves. *Bioscience*, **41**, 331–336.

Clark, R.B. (1992). *Marine pollution* (3rd edn). Clarendon Press, Oxford.

Clarke, A. (1992). Is there a latitudinal diversity clinc in the sea? *Trends in Ecology and Evolution*, **7**, 286–287.

Colman, J. (1933). The nature of the intertidal zonation of plants and animals. *Journal of the Marine Biological Association of the UK*, **18**, 435–476.

Connell, J.H. (1961*a*). The influence of interspecific competition and other factors on the distribution of the barnacle *Chthamalus stellatus*. *Ecology*, **42**, 710–723.

Connell, J.H. (1961*b*). Effects of competition, predation by *Thais lapillus*, and other factors on natural populations of the barnacle *Balanus balanoides*. *Ecological Monographs*, **31**, 61–104.

Connell, J.H. (1970). A predator–prey system in the marine intertidal region. I. *Balanus glandula* and several species of *Thais*. *Ecological Monographs*, **40**, 49–78.

Cook, A., Bamford, O.S., Freeman, J.D.B. and Tedeman, D.J. (1969). A study of homing habits of the limpet. *Animal Behaviour*, **17**, 330–339.

Cook, S.B. (1969). Experiments on homing in the limpet *Siphonaria normalis*. *Animal Behaviour*, **17**, 679–682.

Costlow, J.D. and Tipper, R.C. (1984). *Marine biodeterioration: an interdisciplinary study*. E. and F.N. Spon, London.

Cousens, R. (1986). Quantitative reproduction and reproductive effort by stands of the brown alga *Ascophyllum nodosum* (L.) Le Jolis in southeastern Canada. *Estuarine and Coastal Shelf Science*, **22**, 495–507.

Creese, R. and Underwood, A.J. (1982). Analysis of inter- and intra-specific competition amongst intertidal limpets with different methods of feeding. *Oecologia (Berlin)*, **53**, 337–346.

Crothers, J.H. (1967). The biology of the shore crab *Carcinus maenas* (L.) 1. The background–anatomy, growth and life history. *Field Studies*, **2**, 407–434.

Crothers, J.H. (1968). The biology of the shore crab *Carcinus maenas* (L.) 2. The life of the adult crab. *Field Studies*, **2**, 579–614.

Crothers, J.H. (1985). Dog-whelks: an introduction to the biology of *Nucella lapillus* (L.). *Field Studies*, **6**, 291–360.

Dakin, W.J. (1980). *Australian seashores*. Angus and Robertson, Sydney.

Davenport, J., Gruffydd, Ll.D. and Beaumont, A.R. (1975). An apparatus to supply water of fluctuating salinity and its use in a study of the salinity tolerances of larvae of the scallop *Pecten maximus* L. *Journal of the Marine Biological Association of the UK*, **55**, 391–409.

Dayton, P.K. (1971). Competition, disturbance and community organization: the provision and subsequent utilization of space in a rocky intertidal environment. *Ecological Monographs*, **41**, 351–389.

Dayton, P.K. (1975). Experimental evaluation of ecological dominance in a rocky intertidal algal community. *Ecological Monographs*, **45**, 137–159.

Dayton, P.K. (1985). Ecology of kelp communities. *Annual Review of Ecology and Systematics*, **16**, 215–245.

Denley, E.J. and Underwood, A.J. (1979). Experiments on factors influencing settlement, survival, and growth of two species of barnacles in New South Wales. *Journal of Experimental Marine Biology and Ecology*, **36**, 269–293.

Denny, M.W. (1988). *Biology and mechanics of the wave-swept environment*. Princeton University Press, Princeton, NJ.

Denny, M.W., Daniel, T.L. and Koehl, M.A.R. (1985). Mechanical limits to size in wave-swept organisms. *Ecological Monographs*, **55**, 69–102.

Dethier, M.N. (1984). Disturbance and recovery in intertidal pools: maintenance of mosaic patterns. *Ecological Monographs*, **54**, 99–118.

Dion, P. and Delepine, R. (1983). Experimental ecology of *Gigartina stellata* (Rhodophyta) at Roscoff, France, using an *in situ* culture method. *Botanica Marina*, **26**, 201–211.

Dring, M.J. (1982). *The biology of marine plants*. Cambridge University Press, Cambridge.

Dudgeon, S.R. and Johnson, A.S. (1992). Thick vs. thin: thallus morphology and tissue mechanics influence differential drag and dislodgement of two co-dominant seaweeds. *Journal of Experimental Marine Biology and Ecology*, **165**, 23–43.

Duffy, J.E. (1990). Amphipods on seaweeds: partners or pests? *Oecologia (Berlin)*, **83**, 267–276.

Dumas, J.V. and Witman, J.D. (1993). Predation by herring gulls (*Larus argentatus* Coues (Aves)) on two rocky intertidal crab species (*Carcinus maenas* (L.) and *Cancer irroratus* Say). *Journal of Experimental Marine Biology and Ecology*, **169**, 89–101.

Dye, A.H., Branch, G.M., Castilla, J.C. and Bennett, B.A. (1994). Biological options for the management of the exploitation of intertidal and subtidal resources. In *Rocky shores: exploitation in Chile and South Africa* (ed. W.R. Siegfried), pp. 131–154. Springer-Verlag, Berlin.

Ebling, F.J., Kitching, J.A., Muntz, L. and Taylor, M. (1964). Experimental observations of the destruction of *Mytilus edulis* and *Nucella lapillus* by crabs. *Journal of Animal Ecology*, **33**, 73–82.

Edyvean, R.G.J. and Ford, H. (1986). Population structure of *Lithophyllum incrustans* (Philippi) (Corallinales Rhodophyta) from southwest Wales. *Field Studies*, **6**, 397–405.

Ekaratne, S.U.K. and Crisp, D.J. (1982). Tidal micro-growth bands in intertidal gastropod shells, with an evaluation of band-dating techniques. *Proceedings of the Royal Society of London*, **B 214**, 305–323.

Elner, R.W. and Hughes, R.N. (1978). Energy maximization in the diet of the shore crab *Carcinus maenas*. *Journal of Animal Ecology*, **47**, 103–116.

English Nature (1993). *Conserving England's marine heritage*. English Nature, Peterborough.

Fairweather, P.G. (1988). Predation creates haloes of bare space among prey on rocky seashores in New South Wales. *Australian Journal of Ecology*, **13**, 401–409.

Feare, C.J. and Summers, R.W. (1985). Birds as predators on rocky shores. In *The ecology of rocky shores* (ed. P.G. Moore and R. Seed), pp. 249–264. Hodder and Stoughton, London.

Fenchel, T.M. and Jørgensen, B.B. (1977). Detritus food chains of aquatic ecosystems: the role of bacteria. *Advances in Microbial Ecology*, **1**, 1–58.

Fish, J.D. and Fish, S. (1989). *A student's guide to the seashore*. Unwin Hyman, London.

Foster, M.S. (1990). Organization of macroalgal assemblages in the north east Pacific: the assumption of homogeneity and the illusion of generality. *Hydrobiologia*, **192**, 21–33.

Foster, M.S. and Schiel, D.R. (1988). Kelp communities and sea otters: keystone species or just another brick in the wall? In *The community ecology of sea otters* (ed. G.R. VanBlaricom and J.A. Estes), pp. 92–115. Springer-Verlag, Berlin.

Francis, L. (1976). Social organization within clones of the sea anemone *Anthopleura elegantissima*. *Biological Bulletin of the Marine Biological Laboratory, Woods Hole*, **150**, 361–376.

Gabbott, P.A. and Larman, V.N. (1987). The chemical basis of gregariousness in cirripedes: a review (1953—1984). In *Barnacle biology* (ed. A.J. Southward), pp. 377–388. A.A. Balkema, Rotterdam.

Gaines, S. and Roughgarden, J. (1985). Larval settlement rate: a leading determinant of structure in an ecological community of the marine intertidal zone. *Proceedings of the National Academy of Sciences, USA*, **82**, 3707–3711.

Gappa, J.L., Tablado, A. and Magaldi, N.H. (1990). Influence of sewage pollution on a rocky intertidal community dominated by the mytilid *Brachidontes rodriguezi*. *Marine Ecology Progress Series*, **63**, 163–175.

Gaylord, B., Blanchette, C.A. and Denny, M.W. (1994). Mechanical consequences of size in wave-swept algae. *Ecological Monographs*, **64**, 287–313.

Gendron, R.P. (1977). Habitat selection and migratory behaviour of the intertidal gastropod *Littorina littorea* (L.). *Journal of Animal Ecology*, **46**, 79–92.

Gibb, J. (1956). Food, feeding habits and territory of the rock-pipit, *Anthus spinoletta*. *Ibis*, **98**, 506–530.

Gibbs, P.E. (1993). Phenotypic changes in the progeny of *Nucella lapillus* (Gastropoda) transplanted from an exposed shore to sheltered inlets. *Journal of Molluscan Studies*, **59**, 187–194.

Gibbs, P.E., Pascoe, P.L. and Burt, G.R. (1988). Sex change in the female dog-whelk, *Nucella lapillus*, induced by tributyltin from anti-fouling paints. *Journal of the Marine Biological Association of the UK*, **68**, 715–731.

Gibson, R.N. (1972). The vertical distribution and feeding relationships of intertidal fish on the Atlantic coast of France. *Journal of Animal Ecology*, **41**, 189–207.

Gosner, K.L. (1979). *A field guide to the Atlantic seashore*. Houghton Mifflin, Boston.

Goss-Custard, J.D., Durell, Le V. dit, McGrorty, S., Reading, C.J. and Clarke, R.T. (1981). Factors affecting the occupation of mussel (*Mytilus edulis*) beds by oystercatchers (*Haematopus ostralaegus*) on the Exe estuary, Devon. In *Feeding and survival strategies of estuarine organisms* (ed. N.V. Jones and W.J. Wolff), pp. 217–229. Plenum Press, London.

Goss-Custard, S., Jones, J., Kitching, J.A. and Norton, T.A. (1979). Tide-pools of Carrigathorna and Barloge Creek. *Philosophical Transactions of the Royal Society of London*, **B 287**, 1–44.

Gruet, Y. (1986). Spatio-temporal changes of sabellarian reefs built by the sedentary polychaete *Sabellaria alveolata* (Linne). *Pubblicazioni della Stazione zoologica Napoli (Marine Ecology)*, **7**, 303–319.

Gubbay, S. (1988). *A coastal directory for marine nature conservation*. Marine Conservation Society, Ross-on-Wye.

Gubbay, S. (1989). *Coastal and sea-use management*. Marine Conservation Society, Ross-on-Wye.

Hagen, N.T. and Mann, K.H. (1992). Functional response of the predators American lobster *Homarus americanus* (Milne-Edwards) and Atlantic wolffish *Anarhichas lupus* (L.) to increasing numbers of the green sea urchin *Strongylocentrotus droebachiensis* Müller. *Journal of Experimental Marine Biology and Ecology*, **159**, 89–112.

Hartnoll, R.G. and Hawkins, S.J. (1985). Patchiness and fluctuations on moderately exposed rocky shores. *Ophelia*, **24**, 53–63.

Hawkins, S.J. and Hartnoll, R.G. (1983). Grazing of intertidal algae by marine invertebrates. *Oceanography and Marine Biology Annual Review*, **21**, 195–282.

Hawkins, S.J. and Jones, H.D. (1992). *Rocky shores*. Immel Publishing, London.

Hawkins, S.J., Hartnoll, R.G., Kain, J.M. and Norton, T.A. (1992). Plant–animal interactions on hard substrata in the north-east Atlantic. In *Plant–animal interactions in the marine benthos* (ed. D.M. John, S.J. Hawkins and J.H. Price), pp. 1–32. Clarendon Press, Oxford.

Hay, M.E. and Fenical, W. (1988). Marine plant–herbivore interactions: the ecology of chemical defence. *Annual Review of Ecology and Systematics*, **19**, 111–145.

Hayward, P.J. (1995). *Handbook of the marine fauna of North-West Europe.* Oxford University Press, Oxford.

Hayward, P.J. and Nelson-Smith, A. (1995). *Collins guide to the seashore of Britain and Europe.* Harper Collins, London.

Hayward, P.J. and Ryland, J.S. (1990). *The marine fauna of the British Isles and Northwest Europe*, vols. I and II. Oxford University Press, Oxford.

Hill, A.S. and Hawkins, S.J. (1991). Seasonal and spatial variation of epilithic microalgal distribution and abundance and its ingestion by *Patella vulgata* on a moderately exposed rocky shore. *Journal of the Marine Biological Association of the UK*, **71**, 403–423.

Hopkins, J.T. (1964). A study of the diatoms of the Ouse estuary Sussex III. The seasonal variation in the littoral epiphyte flora and the shore plankton. *Journal of the Marine Biological Association of the UK*, **44**, 613–644.

Huggett, J. and Griffiths, C.L. (1986). Some relationships between elevation, physico-chemical variables and biota of intertidal rock pools. *Marine Ecology Progress Series*, **29**, 189–197.

Hughes, R.G. (1975). The distribution of epizoites on the hydroid *Nemertesia antennina* (L.). *Journal of the Marine Biological Association of the UK*, **55**, 275–294.

Hughes, R.N. and Elner, R.W. (1979). Tactics of a predator, *Carcinus maenas*, and morphological responses of the prey, *Nucella lapillus*. *Journal of Animal Ecology*, **48**, 65–78.

Hunter, R.D. and Russell-Hunter, W.D. (1983). Bioenergetic and community changes in intertidal Aufwuchs grazed by *Littorina littorea*. *Ecology*, **64**, 761–769.

Janke, K. (1990). Biological interactions and their role in the community structure in the rocky intertidal of Helgoland (German Bight, North Sea). *Helgolander Wissenschaftliche Meeresuntersuchungen*, **44**, 219–263.

Johannesson, B. (1986). Shell morphology of *Littorina saxatilis* Olivi: the relative importance of physical factors and predation. *Journal of Experimental Marine Biology and Ecology*, **102**, 183–195.

John, D.M., Hawkins, S.J. and Price, J.H. (eds) (1992). *Plant–animal interactions in the marine benthos.* Clarendon Press, Oxford.

Jones, N.S. (1948). Observations and experiments on the biology of *Patella vulgata* at Port St. Mary, Isle of Man. *Proceedings and Transactions of the Liverpool Biological Society*, **56**, 60–77.

Jones, N.S. and Kain, J.M. (1967). Subtidal algal colonization following the removal of *Echinus. Helgolander Wissenschaftliche Meeresuntersuchungen*, **15**, 460–466.

Jørgensen, C.B. (1966). *The biology of suspension feeding*. Pergamon Press, Oxford.

Kain, J.M. (1979). A view of the genus *Laminaria*. *Oceanography and Marine Biology Annual Review*, **17**, 101–161.

Kendall, M.A. (1988). The age and size structure of some northern populations of the trochid gastropod *Monodonta lineata*. *Journal of Molluscan Studies*, **53**, 213–222.

Kingham, L. (1991). Environmental monitoring at the Sullom Voe oil terminal, Shetland. *Proceedings of the One Day Pollution Conference, 5 September 1991*. Dyfed County Council, Carmarthen.

Kitching, J.A. (1987). Ecological studies at Lough Hyne. *Advances in Ecological Research*, **17**, 115–186.

Kitching, J.A. and Thain, V.M. (1983). The ecological impact of the sea urchin *Paracentrotus lividus* (Lamarck) in Lough Ine, Ireland. *Philosophical Transactions of the Royal Society of London*, **B 300**, 513–552.

Kitching, J.A., Muntz, L. and Ebling, F.J. (1966). The ecology of Lough Ine XV. The ecological significance of shell and body forms in *Nucella*. *Journal of Animal Ecology*, **35**, 113–126.

Knight, M. and Parke, M. (1950). A biological study of *Fucus vesiculosus* and *F. serratus*. *Journal of the Marine Biological Association of the UK*, **29**, 439–514.

Knight-Jones, E.W. (1953). Laboratory experiments on gregariousness during settling in *Balanus balanoides* and other barnacles. *Journal of Experimental Biology*, **30**, 584–598.

Knight-Jones, E.W., Bailey, J.H. and Isaac, M.J. (1971). Choice of algae by larvae of *Spirorbis*, particularly of *Spirorbis spirorbis*. In *Proceedings of the 4th European Marine Biology Symposium* (ed. D.J. Crisp), pp. 89–104. Cambridge University Press, Cambridge.

Koehl, M.A.R. and Alberte, R.S. (1988). Flow, flapping, and photosynthesis of *Nereocystis luetkeana*: a functional comparison of undulate and flat blade morphologies. *Marine Biology*, **99**, 435–444.

LaBarbera, M. (1984). Feeding currents and particle capture mechanisms in suspension feeding animals. *American Zoologist*, **24**, 71–84.

Lewis, J.R. (1955). The mode of occurrence of the universal intertidal zones in Great Britain. *Journal of Ecology*, **43**, 270–290.

Lewis, J.R. (1964). *The ecology of rocky shores*. English Universities Press, London.

Lewis, J.R. (1977). The role of physical and biological factors in the distribution and stability of rocky shore communities. In *Biology of benthic organisms* (ed. B.F. Keegan, P.O. Ceidigh and P.J.S. Boaden), pp. 417–424. Pergamon Press, Oxford.

Little, C. (1990). *The terrestrial invasion. An ecophysiological approach to the origins of land animals*. Cambridge University Press, Cambridge.

Little, C., Morritt, D., Paterson, D.M., Stirling, P. and Williams, G.A. (1990). Preliminary observations on factors affecting foraging activity in the limpet *Patella vulgata*. *Journal of the Marine Biological Association of the UK*, **70**, 181–195.

Little, C., Partridge, J.C. and Teagle, L. (1991). Foraging activity of limpets in normal and abnormal tidal regimes. *Journal of the Marine Biological Association of the UK*, **71**, 537–554.

Littler, M.M. and Murray, S.N. (1975). Impact of sewage on the distribution, abundance and community structure of rocky intertidal macro-organisms. *Marine Biology*, **30**, 277–291.

Lubchenco, J. (1978). Plant species diversity in a marine intertidal community: importance of herbivore food preference and algal competitive abilities. *American Naturalist*, **112**, 23–39.

Lubchenco, J. (1982). Effects of grazers and algal competitors on fucoid colonization in tide-pools. *Journal of Phycology*, **18**, 544–550.

Lubchenco, J. and Gaines, S.D. (1981). A unified approach to marine plant–herbivore interactions. I. Populations and communities. *Annual Review of Ecology and Systematics*, **12**, 405–437.

Lubchenco, J. and Menge, B.A. (1978). Community development and persistence in a low rocky intertidal zone. *Ecological Monographs*, **48**, 67–94.

Lüning, K. (1988). Photoperiodic control of sorus formation in the brown alga *Laminaria saccharina*. *Marine Ecology Progress Series*, **45**, 137–144.

McGaw, I.J. and Naylor, E. (1992). Distribution and rhythmic locomotor patterns of estuarine and open-shore populations of *Carcinus maenas*. *Journal of the Marine Biological Association of the UK*, **72**, 599–609.

Mann, K.H. (1988). Production and use of detritus in various freshwater, estuarine and coastal marine ecosystems. *Limnology and Oceanography*, **33**, 910–930.

Menge, B.A. (1976). Organization of the New England rocky intertidal community: role of predation, competition and environmental heterogeneity. *Ecological Monographs*, **46**, 355–393.

Menge, B.A. (1982). Effects of feeding on the environment: Asteroidea. In *Echinoderm nutrition* (ed. M. Jangoux and J.M. Lawrence), pp. 521–551. A.A. Balkema, Rotterdam.

Menge, B.A. (1992). Community regulation: under what conditions are bottom-up factors important on rocky shores? *Ecology*, **73**, 755–765.

Menge, B.A., Berlow, E.L., Blanchette, C.A., Navarrete, S.A. and Yamada, S.B. (1994). The keystone species concept: variation in interaction strength in a rocky intertidal habitat. *Ecological Monographs*, **64**, 249–286.

Mettam, C. (1994). Intertidal zonation of animals and plants on rocky shores in the Bristol Channel and Severn Estuary–the northern shores. *Biological Journal of the Linnean Society*, **51**, 123–147.

Miller, R.J. (1985). Sea urchin pathogen: a possible tool for biological control. *Marine Ecology Progress Series*, **21**, 169–174.

Miller, R.J., Mann, R.H. and Scarratt, D.J. (1971). Production potential of a seaweed–lobster community in eastern Canada. *Journal of the Fisheries Research Board of Canada*, **28**, 1733–1738.

Moore, H.B. (1934). The biology of *Balanus balanoides*. I. Growth rate and its relation to size, season and tidal level. *Journal of the Marine Biological Association of the UK*, **19**, 851–868.

Moore, P.G. and Seed, R. (eds) (1985). *The ecology of rocky coasts*. Hodder and Stoughton, London.

Morris, S. and Taylor, A.C. (1983). Diurnal and seasonal variation in physico-chemical conditions within intertidal rock pools. *Estuarine and Coastal Shelf Science*, **17**, 339–355.

Morton, J. and Miller, M. (1973). *The New Zealand sea shore* (2nd edn). Collins, London.

Nelson, B.V. and Vance, R.R. (1979). Diel foraging patterns of the sea urchin *Centrostephanus coronatus* as a predator avoidance strategy. *Marine Biology*, **51**, 251–258.

Newell, R.C. (1979). *Biology of intertidal animals* (3rd edn). Marine Ecological Surveys Ltd., Faversham, Kent.

Northcott, S.J., Gibson, R.N. and Morgan, E. (1990). The persistence and modulation of endogenous circatidal rhythmicity in *Lipophrys pholis* (Teleostei). *Journal of the Marine Biological Association of the UK*, **70**, 815–827.

Norton, T.A. (1985). The zonation of seaweeds on rocky shores. In *The ecology of rocky coasts* (ed. P.G. Moore and R. Seed), pp. 7–21. Hodder and Stoughton, London.

Norton, T.A. and Fetter, R. (1981). The settlement of *Sargassum muticum* propagules in stationary and flowing water. *Journal of the Marine Biological Association of the UK*, **61**, 929–940.

Norton, T.A., Ebling, F.J. and Kitching, J.A. (1971). Light and the distribution of organisms in a sea cave. In *Fourth European Marine Biology Symposium* (ed. D.J. Crisp), pp. 409–432. Cambridge University Press, Cambridge.

Norton, T.A., Hawkins, S.J., Manley, N.L., Williams, G.A. and Watson, D.C. (1990). Scraping a living: a review of littorinid grazing. *Hydrobiologia*, **193**, 117–138.

O'Donnell, D. (1991). The role of the Wildlife Service (Office of Public Works) in the management of the Lough Hyne marine nature reserve. In *The ecology of Lough Hyne* (eds A.A. Myers, C. Little, M.J. Costello and J.C. Partridge), pp. 165–168. Royal Irish Academy, Dublin.

Okamura, B. (1986). Formation and disruption of aggregations of *Mytilus edulis* in the fouling community of San Francisco Bay, California. *Marine Ecology Progress Series*, **30**, 275–282.

Okamura, B. (1990). Behavioural plasticity in the suspension feeding of benthic animals. In *Behavioural mechanisms of food selection* (ed. R.N. Hughes), pp. 637–660. Springer-Verlag, Berlin.

Okamura, B. (1992). Microhabitat variation and patterns of growth and feeding in a marine bryozoan. *Ecology*, **73**, 1502–1513.

Ottaway, J.R. (1978). Population ecology of the intertidal anemone *Actinia tenebrosa* I. Pedal locomotion and intraspecific aggression. *Australian Journal of Marine and Freshwater Research*, **29**, 787–802.

Paine, R.T. (1974). Intertidal community structure: experimental studies on the relationship between a dominant competitor and its principal predator. *Oecologia (Berlin)*, **15**, 93–120.

Paine, R.T. (1980). Foodwebs: linkage, interaction strength and community infrastructure. *Journal of Animal Ecology*, **49**, 667–685.

Paine, R.T. (1994). *Marine rocky shores and community ecology: an experimentalist's perspective.* Ecology Institute, Oldendorf/Luhe, Germany.

Paine, R.T. and Levin, S.A. (1981). Intertidal landscapes: disturbance and the dynamics of pattern. *Ecological Monographs*, **51**, 145–178.

Paine, R.T., Castilla, J.C. and Cancino, J. (1985). Perturbation and recovery patterns of starfish dominated intertidal assemblages in Chile, New Zealand and Washington State. *American Naturalist*, **125**, 679–681.

Palmer, A.R. (1990). Effect of crab effluent and scent of damaged conspecifics on feeding, growth, and shell morphology of the Atlantic dog whelk *Nucella lapillus* (L.). *Hydrobiologia*, **193**, 155–182.

Parke, M. (1948). Studies on British Laminariaceae. I. Growth in *Laminaria saccharina* (L.) Lamour. *Journal of the Marine Biological Association of the UK*, **27**, 651–709.

Pechenik, J.A., Eyster, L.S., Widdows, J. and Bayne, B.L. (1990). The influence of food concentration and temperature on growth and morphological differentiation of blue mussel *Mytilus edulis* L. larvae. *Journal of Experimental Marine Biology and Ecology*, **136**, 47–64.

Pedrotti, M.L. (1993). Spatial and temporal distribution and recruitment of echinoderm larvae in the Ligurian Sea. *Journal of the Marine Biological Association of the UK*, **73**, 513–530.

Pethick, J. (1984). *An introduction to coastal geomorphology.* Arnold, London.

Petraitis, P.S. (1990). Direct and indirect effects of predation, herbivory and surface rugosity on mussel recruitment. *Oecologia (Berlin)*, **83**, 405–413.

Quigley, M. and Crump, R. (1986). *Animals and plants of rocky shores.* Blackwell, Oxford.

Quinn, J.F., Wing, S.R. and Botsford, L.W. (1993). Harvest refugia in marine invertebrate fisheries: models and applications to the red sea urchin, *Strongylocentrotus franciscanus. American Zoologist*, **33**, 537–550.

Raffaelli, D. (1982). Recent ecological research on some European species of *Littorina. Journal of Molluscan Studies*, **48**, 342–354.

Reid, D.G. (1993). Barnacle-dwelling ecotypes of three British *Littorina* species and the status of *Littorina neglecta* Bean. *Journal of Molluscan Studies*, **59**, 51–62.

Reimchen, T.E. (1979). Substratum heterogeneity, crypsis and colour polymorphism in an intertidal snail (*Littorina*). *Canadian Journal of Zoology*, **57**, 1070–1085.

Remmert, H. (ed.) (1991). *The mosaic-cycle concept of ecosystems*. Springer-Verlag, Berlin.

Richardson, C.A. (1989). An analysis of the microgrowth bands in the shell of the common mussel *Mytilus edulis*. *Journal of the Marine Biological Association of the UK*, **69**, 477–491.

Ricketts, E.F. and Calvin, J. (1952). *Between Pacific tides* (3rd edn). Stanford University Press, Stanford, California.

Rowley, R.J. (1989). Settlement and recruitment of sea urchin (*Strongylocentrotus* spp.) in a sea-urchin barren ground and a kelp bed: are populations regulated by settlement or post-settlement processes? *Marine Biology*, **100**, 483–494.

Santelices, B. (1990a). Patterns of organization of intertidal and shallow sublittoral vegetation in wave exposed habitats of central Chile. *Hydrobiologia*, **192**, 33–57.

Santelices, B. (1990b). Patterns of reproduction, dispersal and recruitment in seaweeds. *Oceanography and Marine Biology Annual Review*, **28**, 177–276.

Savy, S. (1987). Activity pattern of the sea-star *Marthasterias glacialis* in Port-Cros Bay (France, Mediterranean coast). *Pubblicazioni della Stazione zoologica Napoli (Marine Ecology)*, **8**, 97–106.

Scagel, R.F., Bandoni, R.J., Maze, J.R., Rouse, G.E., Schofield, W.B. and Stein, J.R. (1984). *Plants. An evolutionary survey*. Wadsworth Publishing, Belmont, California.

Schiel, D.R. and Foster, M.S. (1986). The structure of subtidal algal stands in temperate waters. *Oceanography and Marine Biology Annual Review*, **24**, 265–307.

Schonbeck, M.W. and Norton, T.A. (1978). Factors controlling the upper limits of fucoid algae on the shore. *Journal of Experimental Marine Biology and Ecology*, **31**, 303–313.

Schonbeck, M.W. and Norton, T.A. (1979). Drought-hardening in the upper-shore seaweeds *Fucus spiralis* and *Pelvetia canaliculata*. *Journal of Ecology*, **67**, 687–696.

Schonbeck, M.W. and Norton, T.A. (1980). Factors controlling the lower limits of fucoid algae on the shore. *Journal of Experimental Marine Biology and Ecology*, **43**, 131–150.

Seed, R. (1969). The ecology of *Mytilus edulis* L. (Lamellibranchiata) on exposed rocky shores. Part 1: breeding and settlement. *Oecologia (Berlin)*, **3**, 277–316.

Seed, R. (1974). Morphological variations in *Mytilus* from the Irish coasts in relation to the occurrence and distribution of *M. galloprovincialis* Link. *Cahiers de Biologie Marine*, **15**, 1–25.

Seed, R. (1985). Ecological pattern in the epifaunal communities of coastal macroalgae. In *The ecology of rocky coasts* (eds P.G. Moore and R. Seed), pp. 22–35. Hodder and Stoughton, London.

Shick, J.M. (1991). *A functional biology of sea anemones.* Chapman & Hall, London.

Siegfried, W.R. (ed.) (1994). *Rocky shores: exploitation in Chile and South Africa.* Springer-Verlag, Berlin.

Sih, A., Crowley, P., McPeek, M., Petranka, J. and Strohmeier, K. (1985). Predation, competition, and prey communities: a review of field experiments. *Annual Review of Ecology and Systematics*, **16**, 269–311.

Sloan, N.A. (1980). Aspects of feeding biology of asteroids. *Oceanography and Marine Biology Annual Review*, **18**, 57–124.

Smith, S.D.A. and Simpson, R.D. (1992). Monitoring the shallow sublittoral using the fauna of kelp (*Ecklonia radiata*) holdfasts. *Marine Pollution Bulletin*, **24**, 46–52.

Sousa, W.P. (1979). Experimental investigations of disturbance and ecological succession in a rocky intertidal algal community. *Ecological Monographs*, **49**, 227–254.

Southward, A.J. (1958). Note on the temperature tolerance of some intertidal animals in relation to environmental temperatures and geographical distribution. *Journal of the Marine Biological Association of the UK*, **37**, 49–66.

Southward, A.J. (1991). Forty years of changes in species composition and population density of barnacles on a rocky shore near Plymouth. *Journal of the Marine Biological Association of the UK*, **71**, 495–513.

Southward, A.J. and Southward, E.C. (1978). Recolonisation of rocky shores in Cornwall after use of toxic dispersants to clean up the Torrey Canyon oil spill. *Journal of the Fisheries Research Board of Canada*, **35**, 682–706.

Stebbing, A.R.D. (1973). Competition for space between the epiphytes of *Fucus serratus* L. *Journal of the Marine Biological Association of the UK*, **53**, 247–261.

Steneck, R.S. (1986). The ecology of coralline algal crusts: convergent patterns and adaptive strategies. *Annual Review of Ecology and Systematics*, **17**, 273–303.

Stephenson, T.A. and Stephenson, A. (1972). *Life between tidemarks on rocky shores.* W.H. Freeman, San Francisco.

Stimson, J. (1970). Territorial behavior of the owl limpet, *Lottia gigantea. Ecology*, **51**, 113–118.

Suchanek, T.H. (1993). Oil impacts on marine invertebrate populations and communities. *American Zoologist*, **33**, 510–523.

Thain, V.M. (1971). Diurnal rhythms in snails and starfish. In *Fourth European Marine Biology Symposium* (ed. D.J. Crisp), pp. 513–537. Cambridge University Press, Cambridge.

Thain, V.M., Thain, J.F. and Kitching, J.A. (1985). Return of the prosobranch *Gibbula umbilicalis* (da Costa) to the littoral region after displacement to the shallow sublittoral. *Journal of Molluscan Studies*, **51**, 205–210.

Thomas, M.L.H. (1986). A physically derived exposure index for marine shorelines. *Ophelia*, **25**, 1–13.

Thomas, M.L.H. (1994). Littoral communities and zonation on rocky shores in the Bay of Fundy, Canada: an area of high tidal range. *Biological Journal of the Linnean Society*, **51**, 149–168.

Thompson, T.E. (1976). *Biology of opisthobranch molluscs*. Vol. I. Ray Society, London.

Thompson, T.E. and Brown, G.H. (1984). *Biology of opisthobranch molluscs*. Vol. II. Ray Society, London.

Todd, C.D. (1981). The ecology of nudibranch molluscs. *Oceanography and Marine Biology Annual Review*, **19**, 141–234.

Underwood, A.J. (1972). Tide model analysis of the zonation of intertidal prosobranchs: I. Four species of *Littorina* (L.). *Journal of Experimental Marine Biology and Ecology*, **9**, 239–255.

Underwood, A.J. (1976). Food competition between age-classes in the intertidal neritacean *Nerita atramentosa* Reeve (Gastropoda: Prosobranchia). *Journal of Experimental Marine Biology and Ecology*, **23**, 145–154.

Underwood, A.J. (1978). A refutation of critical tidal levels as determinants of the structure of intertidal communities on British shores. *Journal of Experimental Marine Biology and Ecology*, **33**, 261–276.

Underwood, A.J. (1985). Physical factors and biological interactions: the necessity and nature of ecological experiments. In *The ecology of rocky coasts* (ed. P.G. Moore and R. Seed), pp. 372–390. Hodder and Stoughton, London.

Underwood, A.J. (1986). The analysis of competition by field experiments. In *Community ecology: pattern and process* (ed. J. Kikkawa and D.J. Anderson), pp. 240–268. Blackwell, London.

Underwood, A.J. (1991). The logic of ecological experiments: a case history from studies of the distribution of macro-algae on rocky intertidal shores. *Journal of the Marine Biological Association of the UK*, **71**, 841–866.

Underwood, A.J. and Fairweather, P.G. (1989). Supply-side ecology and benthic marine assemblages. *Trends in Ecology and Evolution*, **4**, 16–20.

Underwood, A.J. and Kennelly, S.J. (1990). Ecology of marine algae on rocky shores and subtidal reefs in temperate Australia. *Hydrobiologia*, **192**, 3–20.

Underwood, A.J., Denley, E.J. and Moran, M.J. (1983). Experimental analysis of the structure and dynamics of mid-shore rocky intertidal communities in New South Wales. *Oecologia (Berlin)*, **56**, 202–219.

Van Alstyne, K.L. (1989). Adventitious branching as a herbivore-induced defense in the intertidal brown alga *Fucus distichus*. *Marine Ecology Progress Series*, **56**, 169–176.

Van Blaricom, G.R. and Estes, J.A. (eds) (1988). *The community ecology of sea otters*. Springer-Verlag, Berlin.

Viejo, R.M. and Arrontes, J. (1992). Interactions between mesograzers inhabiting *Fucus vesiculosus* in northern Spain. *Journal of Experimental Marine Biology and Ecology*, **162**, 97–111.

Vogt, H. and Schramm, W. (1991). Conspicuous decline of *Fucus* in Kiel Bay (Western Baltic): what are the causes? *Marine Ecology Progress Series*, **69**, 189–194.

Voipio, A. (ed.) (1981). *The Baltic sea*. Elsevier, Amsterdam.

Ward, R.D. (1990). Biochemical genetic variation in the genus *Littorina* (Prosobranchia: Mollusca). *Hydrobiologia*, **193**, 53–69.

Warwick, R.M. and Clarke, K.R. (1991). A comparison of some methods for analysing changes in benthic community structure. *Journal of the Marine Biological Association of the UK*, **71**, 225–244.

Williams, G.A. (1992). The effect of predation on the life histories of *Littorina obtusata* and *Littorina mariae*. *Journal of the Marine Biological Association of the UK*, **72**, 403–416.

Williams, G.B. (1964). The effect of extracts of *Fucus serratus* in promoting the settlement of larvae of *Spirorbis borealis* (Polychaeta). *Journal of the Marine Biological Association of the UK*, **44**, 397–414.

Yonge, C.M. (1949). *The sea shore*. Collins, London.

Zach, R. (1978). Selection and dropping of whelks by north-western crows. *Behaviour*, **67**, 134–148.

Index

(Bold numbers denote reference to illustrations)